# WOUNDED WORKERS:
# THE POLITICS OF MUSCULOSKELETAL INJURIES

One-third of serious workplace injuries in the 1990s can be classified as musculoskeletal injuries (MSIs). Painful and potentially permanent, MSIs are often difficult to treat. Despite evidence that applied ergonomics and early intervention can protect workers from these injuries, the medical, social, legal, and political responses to MSIs have been ambiguous or even hostile. For example, even as the number of work-related MSIs increases, many provinces and states are revising Workers' Compensation regulations to exclude these injuries from coverage.

MSIs of the hands, arms, neck, and shoulders strike women twice as often as men. Among the reasons for this difference: women work in jobs that tend to be tedious and repetitive (such as data input, electronics assembly, cashiering, sewing, and poultry packing) and that are often not protected by collective bargaining; most workstations are designed for a so-called average male body, which forces women to reach awkwardly; and domestic duties usually prevent women from resting their hands after the paid work is done. In addition, since women's jobs are usually perceived as 'light work,' Workers' Compensation claims are often disputed by employers. Even when a claim is accepted, Workers' Compensation pays only a fraction of women's already low wages.

A person who develops an MSI faces a labyrinth of bureaucracy. *Wounded Workers* maps out the current situation for patients, caregivers, and advocates in Canada and the United States. It reviews current therapies (mainstream and alternative), ergonomics, legislation, litigation, union–management relations, and the trend towards de-compensation. Most MSI books focus on medical self-help. *Wounded Workers* exposes the hidden policy agendas that face MSI patients and their caregivers, and points to further resources for overcoming the barriers to recovery.

PENNEY KOME is an author and award-winning journalist living in Calgary, Alberta.

PENNEY KOME

# Wounded Workers: The Politics of Musculoskeletal Injuries

UNIVERSITY OF TORONTO PRESS
Toronto Buffalo London

© Penney Kome 1998
Published by University of Toronto Press Incorporated
Toronto  Buffalo  London
Printed in Canada

ISBN 0-8020-0741-4 (cloth)
ISBN 0-8020-7795-1 (paper)

Printed on acid-free paper

---

**Canadian Cataloguing in Publication Data**

Kome, Penney
  Wounded workers: the politics of musculoskeletal injuries

  Includes bibliographical references and index.
  ISBN 0-8020-0741-4 (bound)   ISBN 0-8020-7795-1 (pbk.)

  1. Overuse injuries. 2. Industrial hygiene. I. Title.

  RD97.5.K65 1998      363.11     C98-930822-7

---

University of Toronto Press acknowledges the financial assistance to its publishing program of the Canada Council for the Arts and the Ontario Arts Council.

This book is dedicated to
the memory of my father,

      Hal Kome,

who lived to write,
and wrote for a living – in longhand.

2 March 1924 – 2 May 1996

# Contents

ACKNOWLEDGMENTS    ix
INTRODUCTION    xi

1  MSIs, RSIs, and the Workplace    3
2  Why There Are No RSIs in the Land of Oz (They Go by Other Names)    23
3  Waiting Rooms    30
4  Bigger Than a Breadbox?    62
5  Compensation? But You Don't Look Disabled    71
6  So Sue Me    100
7  Staring at the Screen: Computer Workstations    105
8  Fitting the Jobs to the Workers    118
9  Beyond Grieving    142
10  Legislation in Other Jurisdictions    165
11  The Battle over Legislation    177
12  By the Fingernails    190

NOTES    207
SELECTED BIBLIOGRAPHY    231
APPENDIX    237
INDEX    241

# Acknowledgments

This book draws from many sources, but perhaps the richest source of all has been the SOREHAND mailing list on the Internet (http://www.ucsf.edu/sorehand). The approximately 750 list members (at any given time) have added insight and dimension beyond anything found in journals. Several have become friends as well as cyberpals.

Many thanks also to PJ Edington of the Center for Office Technology and David Felinski of the American Automobile Manufacturers Association, who, in collaboration with Hank Lick of Ford Motor Company, put together the most extraordinarily stimulating and enriched four-day conference on policy issues that I have ever attended.

Marilyn Segal provided invaluable leads and encouragement from the very beginning. Lois Weninger and Lynn Bueckert's efforts at the Women and Work Research and Education Society first piqued my interest in this subject. Wendy King and David Cohen provided background and publications. Barbara Silverstein gave me her undivided attention for two full hours on a day that must have been extremely tense for her.

On a personal note, my thanks go to Bob, Sanford, and Graham for their understanding and support, and mainly for taking all those camping trips and ski trips that left me time to work.

A great many others have contributed to this book, either directly or indirectly. Readers who take the time to follow up the works cited in the footnotes will be amply rewarded by meeting the scintillating minds behind the documents and interviews.

PENNEY KOME

# Introduction

WHY THIS BOOK?

Most books dealing with work-related musculoskeletal injuries (MSIs) are self-help medical books. This one isn't. That is, there's a short self-help section about medical care, but the main text of the book addresses politics, both small-p and big-P (partisan) politics.

MSIs, this book will argue, push patients right into the intersection of several political dynamics – all of which affect not only the kind of healthcare patients receive, but also the way they are treated at their jobs, by their friends and families, and by their governments.

A person who seeks help for aching wrists, elbows, shoulders, or other upper-extremity pain and dysfunction, encounters politics:

- in the workplace, in the form of gender distinctions, class distinctions, and the ongoing thrust towards downsizing and greater productivity per worker;
- in healthcare, in the form of gender distinctions, lack of precision in occupational health issues, health management organizations (HMOs) or capitation fee schedules that are geared towards acute cases rather than chronic conditions;
- professional conflicts about whether or how to treat MSIs, and the friction between established medical practices and alternative health care;

- in dealing with Workers' Compensation Boards (WCBs), which were established as no-fault insurance plans but which have been breaking down under increased caseloads and determined assaults from local and national employers set on cutting benefits and medical payments in order to reduce Workers' Compensation (WC) costs;
- in litigation, in suing WCBs or in bringing product liability suits against equipment manufacturers;
- in ergonomics, a developing field without clear standards or qualifications, hampered by internal differences;
- in labour/management relations, where ergonomic changes and WC claims may become bargaining chips; and
- in legislation, where the trend towards deregulation runs counter to what patients may see as an urgent need, first, to recognize MSIs as work related and second, to enact ergonomics regulations to protect workers from injuries.

A patient can hardly escape being affected and may in fact be trapped by these politics, without ever clearly perceiving what obstacles she's encountering. And the pronoun 'she' applies to two-thirds of patients.

---

*About Gender*

According to the booklet *Occupational Repetitive Strain Injuries* published by Alberta Occupational Health and Safety, 'studies have shown women to experience higher rates of RSI [repetitive strain injury] than men. This is most likely due to the large number of women employed in those industries (e.g., clothing manufacturing, electrical assembly, food processing, clerical workers) which involve highly repetitive work or other ... RSI factors.'[1]

---

Women report two-thirds of upper-extremity MSIs (and one-third of lower-back pain). Women are concentrated in the kinds of jobs (such as data entry, electronics assembly, poultry packing, cashiering, and sewing) that are statistically at highest risk for

MSIs. When women seek medical help, they often report that doctors suggest the pain is psychosomatic or related to hormones. Currently, standard medical practice is to order the MSI patient to 'rest' – an impossible prescription for women with domestic duties, and especially for single parents. Carpal tunnel syndrome (CTS) has become so common among clerical workers that at least one women's magazine (*Woman's Day*) regularly runs brief tips on how to manage CTS and how soon to return to work after surgery.

Workers' Compensation legislation in some jurisdictions specifically excludes the kinds of clerical jobs in banking and finance where women are concentrated. Even where they're eligible, women's low wages mean they can usually expect lower compensation from WC or from litigation.

Our built environment since the Second World War has been designed to anthropometric standards – literally, measured to fit men. Machinery, equipment, and furniture are all designed for the so-called average man, resulting in desks, chairs, conveyor belts, and other workstations that are too tall for most women. To this day, few ergonomists are women, and the ergonomists who come from the engineering side (especially) don't always understand what's involved with the jobs women do.

Unionized occupations are increasingly likely to include ergonomics in their collective agreements, but women are less likely to work in unionized workplaces than men. As early as 1981, women's organizations tried to organize office workers around computer-safety issues, with little support from major unions.[2] Although some unions have insisted on ergonomic protections in their own workplaces, it was only in mid-1996 that the AFL–CIO launched a national campaign demanding ergonomics regulation for all workplaces – a campaign that explicitly recognizes that women workers are at greater risk for MSIs than men.

And so it goes. Women's inequality in the workforce and in the home is a significant factor, both in the risk that any particular woman will develop an MSI, and in what kind of help she will find to recover from it. Of all the political forces we examine here, the gender dynamic appears in every aspect – except perhaps legislation, where (although there may be unequal impact from a lack of health and safety regulations) the overwhelming force in play seems to be corporate power.

## WHY MSIs? WHY NOW?

MSIs only seem to be modern phenomena. Actually, MSI symptoms are depicted in Egyptian and Greek manuscripts; Bernardo Ramazzini described MSIs in the first compendium of occupational diseases (published in 1713), and medical journals going back to the 1830s have discussions about which MSIs are linked to which occupations.[3]

For some reason, computer injuries have gotten the majority of public attention. Perhaps the notion that innocuous-seeming keyboards and mouses can actually hurt people seems so bizarre that it has captured the public imagination. In fact, while MSIs are closely associated with automation, so far computer-related injuries account for a minority of Workers' Compensation claims for MSIs in the workforce. Unfortunately, as we shall see, the question of compensability for job-related injuries has complicated the medical issues.

## THE BOTTOM LINE

The corporate focus of the 1990s on increasing profits has meant that more and more goods and services are being produced, but by fewer workers. Such a high level of productivity has costs. One such cost is the high unemployment rate: extraordinary numbers of skilled people are without paid work. Another cost, perhaps, is the soaring rate of MSIs.

Literature from the medical and ergonomics fields often includes comments such as this one, from *Occupational Medicine: State of the Art Reviews:* 'The employees most frequently injured,' observed Linda H. Morse, MD, and Lynn Hinds, ANP, 'are those who do not take their coffee or lunch breaks and often work overtime. Although management, particularly in the current economic climate, encourages such practices, they can result in injury to the most valuable and dedicated employees.'[4]

Research into MSIs is just coming into focus. Employers and workers, doctors and ergonomists, unions and WCBs have yet to reach agreement on most issues. Yet a couple of conclusions are already clear.

First, workers need frequent breaks, as often as every half-hour, to prevent injury. Factories or workplaces where supervisors push workers to be productive every minute they're on the line or at their desks have been finding themselves paying out fortunes in Workers' Compensation fines. (Perhaps that's one reason that many employers deny that MSIs exist or are work related.)

Second, humans simply cannot work as long or as hard or as fast as machines. A computer may be able to type around the clock; a person cannot maintain that pace for very long without adverse consequences. The model created by Charles Babbage in 1832, of breaking down every job into its simplest components and then hiring separate workers to perform those components over and over all day – at the cheapest possible wage for the simplest task[5] – has been pushed to its breaking point. In fact, many experts have concluded that more than two hours a day at any one assembly-line job, or more than four hours a day typing at a computer, even with frequent breaks, increases the risk for MSIs.

The bottom line is that the business-led push to deregulate and to downplay risk assessment leaves all members of industrialized society with a choice: we can close our eyes and allow employers to expose workers to hazardous situations – trusting to their enlightened self-interest to protect vulnerable workers, who are often their most skilled and productive employees. Or we can have a huge social dialogue about redefining what kinds of activities constitute bona fide work, how many hours per day or per week individuals should expect to devote to employment, and what kind of living standard the industrialized world can afford to provide for productive adults. If the world is to be divided between the overworked and the unemployed, there are going to be an awful lot of broken bodies around.

# WOUNDED WORKERS

# 1
# MSIs, RSIs, and the Workplace

In the 1990s, more and more people are coming home from work with aching hands, arms, shoulders, and necks. Some of them hurt so much that they seek medical help. Those who get a diagnosis of repetitive strain injury (RSI) – or one of the specific diagnoses that falls under the RSI category – find themselves in a multidisciplinary wonderland of controversy and contradictory claims. Work-related musculoskeletal injuries (MSIs) involve often-conflicting definitions and analyses in several fields, including medical and alternative healthcare, ergonomics, kinesiology and body mechanics, Workers' Compensation and labour litigation, union–management negotiations, and legislation and political lobbying.

TALKING ABOUT ... WHADDYA CALL IT

Information about MSIs comes from several different fields. As a result, the research material behind this book also comes from various fields and from different countries. Just to add to the confusion, there's a variety of terms to describe RSIs, depending on the discipline and country of origin. The box that follows gives a brief glossary.

> *Definitions*
>
> - *Repetitive strain injury* (RSI) is perhaps the most widely used name for a whole group of painful disorders charac-

terized by changes in the soft tissues, including tendons and tendon sheaths, ligaments, nerves, and blood vessels. As used in Canada and the United States, the term RSI usually refers to pain in the upper extremities – the neck, shoulders, arms, and hands. Sometimes the term appears as *repetitive stress, repetitive motion* or *repetitive bodily motion injury* or *disorder.*
- *Cumulative trauma disorder* (or CTD) is an official term commonly used in the United States, where the condition is also deemed to include lower-back pain that is not caused by a single traumatic incident.
- *Occupational overuse syndrome* (OOS) may be synonomous with CTD but *overuse syndrome* by itself often refers to sports injuries.
- *Musculoskeletal injuries* or MSIs (or MSKs) – also known as *work-related upper-extremity* or *limb disorders* (WRUED or WRULD). In Australia, the condition was first described as *cervo-brachial disorder*, which simply indicates the general location of the physical problem; eventually it gave way to the terms *occupational overuse syndrome* or *regional pain syndrome.*
- *Myofascial pain syndrome* may be used by some healthcare practitioners to describe MSIs.
- *Ergonomics-related disorder* – a term that is appearing in proposed legislation and regulations because correcting workplace ergonomics is intended to protect workers from unnecessary strains and sprains.

Some practitioners object to all the terms in this box on the grounds that they describe a theoretical cause for injury – i.e., that too much repetition, especially in awkward body postures, causes microtraumas, which in turn cause chronic conditions – and that these terms do not describe the condition per se. So the same disorders sometimes appear in medical and ergonomics literature as musculoskeletal injuries or MSIs. This is the term used most often in this book, except where text is quoted from other sources.

## WHAT'S HURTING WORKERS' HANDS

Specific diagnoses may include tendinitis, carpal tunnel syndrome, epicondylitis (tennis elbow), thoracic outlet syndrome, frozen shoulder, Reynaud's syndrome, De Quervain's disease, double crush syndrome – the list grows lengthier every year, and patients often have multiple disorders by the time they seek help.

Generally, MSIs involve change to soft tissues that develops over a period of time – usually months or years before a person feels pain or notices clumsiness. Sometimes, however, new computer or automated equipment or a slightly different way of doing things can bring on a local epidemic, and a whole group of workers will be hurt in a short period of time. That's when, in the words of one MSI patient, 'It can get really awful, really fast.'[1]

Many clinicians describe MSIs as progressive. In fact, a scale of injury having as many as five stages has been suggested, but a simpler description offers three distinct stages. In Stage I, upper limbs hurt while the person is at work or is engaged in specific activities. At State II, the aches and pains go home with the worker but disappear overnight. Stage III is the 'crash' stage: pain is constant, painkillers are only partly effective, sleeping patterns are disturbed, and upper limbs (especially hands) just don't function properly.

When a person seeks treatment promptly at the first signs of malfunction (Stage I), MSIs can be corrected without after-effects. Left unchecked, Stage III MSIs can deplete muscle strength so that patients cannot turn door handles or faucets, spread butter on bread, drive, hug, carry groceries, or perform other quotidian tasks.

## WRITER'S CRAMP AND OTHER HISTORICAL MSIs

Medical practitioners have been aware for centuries of the potential for MSIs. When Italian physician and philosopher Benardino Ramazzini published the first text on occupational health, *De Morbis Artificum* (Diseases of Workers), in 1713, he linked awkward postures and forceful actions in the workplace to certain kinds of illness and injury. In a chapter devoted to writer's cramp, he wrote that 'incessant driving of the pen over paper causes intense fatigue of the hand and the whole arm because of the continuous and almost tonic strain on the muscles and tendons, which in course of time results in failure of power in the right hand.'[2] But at the time, writers were a small proportion of the population.

In *Occupation and Disease*, Professor Allard E. Dembe of the University of Massachusetts provides extensive historical research showing that the link between jobs and MSIs has been established for more than a century. MSIs of the hand and wrist first appeared en masse when the Industrial Revolution created a clerical class. From about 1830 on, doctors noted the frequency of writer's cramp, which some attributed to the new technology – of the steel-nib pen. Nineteenth-century medical journals reported some forty varieties of occupational hand disorders, all named after the jobs in which they occurred: milker's cramp, pianist's cramp, sewing spasm, and so on.

In 1875, the first case study of telegrapher's cramp was reported. It signalled the start of a whole new hand–wrist CTD epidemic. A 1911 study found some 64 per cent of 8,153 telegraphers were affected. In 1908, Great Britain recognized telegrapher's cramp as a compensable disorder under the new Workers' Compensation (WC) laws. Dembe remarks on the similarities between telegraph keys and computer mouses in shape, size, and required grip and action.[3]

Today, more than half of all workers in the industrialized workforce process information rather than produce products. Most paid workers are teamed in some way with machines, such as computers, cash registers, assembly lines, cars or trucks, construction equipment, or conveyor belts. There also has been a tendency for paid work to be deskilled – that is, reduced to separate, simple tasks that are repeated over and over – and sped up, in the endless quest for greater productivity.

After having faded from public awareness, suddenly the number of MSI cases has skyrocketed. Only twenty-five years ago, MSIs were seen primarily in athletes and performing artists such as musicians and dancers. Since the mid-1980s, however, MSIs have become the number one cause of time-lost claims for Workers' Compensation. They've been dubbed 'the occupational disease of the information age.'

LET US COUNT THE RATE

We've all seen the emblems of MSIs, perhaps without recognizing what they were. Wrist braces have become almost as common as personal stereos – more common in Silicon Valley areas. When Barbara Janeway was manager of a Durham, North Carolina, bookstore near Research Triangle Park, her training as an occupational therapist cued her to notice wrist braces. 'I'd say that one in twenty people who comes into the store is wearing a wrist brace,' was her estimate.

'The number of ... cumulative-trauma cases reported by workers increased nearly fourfold from 1985 to 1989,' according to a May 1991 article in *Scientific American*.[4] By 1995, according to the *CTD Newsletter*'s private survey, one out of eight workers in the United States was affected by a CTD (which includes lower-back strain).[5] This figure includes only workers. Trina Semorile of the City University of New York estimates that the rate may be as high as one in five 'when you include students and kids.' Teenagers and young adults are showing up in Internet newsgroups and mailing lists for RSI discussions.

In British Columbia, the Workers' Compensation Board (WCB) reported that 'almost one-third of the workers' compensation claims result from ergonomically-related injuries and diseases. In the five year period from 1988 to 1992, the WCB paid out over $400 million for more than 100 thousand ergonomics-related claims. Over 4 million days of work time were lost. This parallels the experience of other jurisdictions such as the U.S., the U.K., Europe and Australia.'[6]

The Washington State Ergonomics Program Guideline reported that

In 1993 more than half of the compensable Washington State Fund workers' compensation claims were attributable to work-related musculoskeletal disorders such as sprains, strains, nerve compression and joint inflammation.

Medical payments and time-loss for these injuries and illnesses cost State Fund employers more than $100 million. In addition, the indirect costs ... can include investigation time, decreased production, training and hiring replacement workers ... [and] are usually estimated to be anywhere from 2–5 times the direct claim costs.[7]

The AFL–CIO has issued a flyer that states, 'Crippling sprain and strain injuries like carpal tunnel syndrome are the #1 job safety problem in the workplace today.'[8]

Epidemiologist Barbara Silverstein puts the numbers into perspective graphically. She has calculated that if all the U.S. workers affected with MSIs stood fingertip to fingertip, they could make a chain of injured hands (and backs) across America, all the way from the East Coast to the West Coast.[9]

BY WAY OF ANALOGY

Ear protectors are now a common sight in workplaces from the factory floor to public parks staff. Science has established, employers have accepted, and the public has come to understand the link between incessant aural assaults and hearing damage. Yet – odd as it now seems – the U.S. military only began to require protective gear for soldiers in very noisy workplaces in 1956. Although noise-induced hearing loss had been noticed by occupational physicians for decades – dubbed 'boilermaker's deafness' in the nineteenth century[10] – it was only after the Second World War that military medics developed documentation on hearing loss among veterans who had been at the front, or who had worked with airplanes and other heavy, noisy machinery.

Perhaps the relationship to working conditions seems obvious in hindsight. Still, hearing loss is a slow, cumulative process that might easily be blamed on aging, or genetics, or personal habits, such as listening to loud music. Moreover, hearing loss is invisible. No employer is likely to watch an employee at work and think, 'Hmm, that person is looking deaf lately.'

The process of proving the workplace connection has been long and painstaking, and the task of protecting workers is far from finished. The National Institute for Occupational Safety and Health (NIOSH) estimates that thirty million U.S. workers are still lacking in protective gear.

MSIs, like hearing loss, are invisible, cumulative, and affected by personal as well as work factors. As with hearing loss, MSIs destroy careers and can cause profound impairment and disability. North American society has finally acknowledged the tragedy of hearing loss, but most workers still have not heard about MSIs. Or they have heard, but they do not believe that they are at risk.

GOOD NEWS, BAD NEWS

The good news is that MSIs are preventable – not easily, but it can be done. Workers and management must work together to identify hazards and modify both workers' habits and the work environment. For instance, posture is a critical issue, and posture is determined by job demands, by the workstation set-up, and by the workers' own dispositions.

The bad news is that the 1990s have seen battle after battle over proposed legislation and regulation aimed at preventing MSIs. In 1995, the U.S. Congress specifically forbade the Occupational Safety and Health Administration (OSHA) from spending any money to research or publish legislation and regulations already drafted and on the table in 1994.[11]

Also in 1995, the newly elected Progressive Conservative government of Ontario moved to strike 'repetitive strain injuries' from the Workers' Compensation Board list of compensable illnesses and to dismantle the bipartite (union and management) Workers' Health and Safety Agency, which had been training workers to recognize ergonomics hazards.[12]

California has seen struggles between the legislature and the Standards Board. Ergonomic protection legislation has been ordered, proposed, watered down, challenged, refused, and reinstated. At this writing, the legislation has been enacted but is back before the courts. In 1995 the California Federation of Labor had to sue the state just to get legislation back into consideration.[13] In British Columbia, draft ergonomics regulations almost made it to proclamation in 1994. They were shelved after employers showed massive opposition during public hearings. Eventually, ergonomics requirements reappeared in the revised British Columbia Occupational Health and Safety Regulations, effective 15 April 1998,[14] as part of the 'General Conditions' section.

There has been a decades-long drive to reduce Workers' Compensation premiums (by reducing injured workers' benefits) across the United States. A similar course has also been pursued in Canada, first by Frank McKenna's government in New Brunswick and most recently by Mike Harris's Ontario government. Unfortunately, the MSI hazard has risen at the very time that workers' protections have been deliberately diminished.

PAIN IS PART OF THE JOB

People who work on fishing boats or in logging or construction are well aware that they'll be using their muscles to earn their living. Many choose such jobs precisely because they enjoy physical activities more than being stuck in an office or a factory. For them, aches and pains come with the territory, and their work usually keeps them at a high level of fitness.

Aching muscles are usually unexpected by people in jobs that are perceived as 'light work,' such as electronic assembly-line workers. Yet light-assembly workers are especially at risk for MSIs. So are a wide range of other occupations, including meat packers, construction workers, hairdressers, musicians, cashiers, bank clerks, fishers, housekeepers, carpenters, postal workers, sewers and cutters, operating room personnel, and huge numbers of computer operators.[15]

Office workers are particularly likely to ignore the early signals that warn of impending musculoskeletal problems, simply because they don't expect their work to be injurious. They rub their aching forearms, take analgesics, and try to carry on until they reach that terrifying stage where they just can't do anything. They may not connect the pain with their jobs at all. Even more mystifying is the gradual onset of back pain – now regarded as another form of MSI – that is related to sitting in one position all day.

BLINDSIDED BY MSI

Women who have pushed themselves through the early stages of MSI to what some call 'the crash' try to warn their colleagues. They squirm when they overhear other women talking about symptoms – what MSI patients recognize as symptoms – apparently without realizing their significance: the arm that aches at night, the recurring numbness in hands or fingers. Women who have 'crashed' also recognize how easy it is for a worker to push herself, to do just one more task or chore, and *then* perhaps take a break. They know how debilitating and frightening Stage III is, to lose function in their hands, to be in so much pain that painkillers don't work. They know the frustration of spending time and energy in trying to find a cure or at least some relief from their condition. And they know what it feels like to have colleagues and superiors minimize or dismiss their complaints with raised eyebrows and a 'But you don't look disabled.'

The last step into severe MSIs can be a nasty shock. Most people have one defining moment – often first thing in the morning – when they realize that their hands are non-functional.

Laboratory technologist May R. ignored the pain building in her neck and shoulders until one day she woke up to find that, 'I couldn't move. My right shoulder and arm were extremely painful and completely useless.' Her doctor's diagnosis: thoracic outlet syndrome. The thoracic surgeon recommended that she have her top two ribs removed to relieve the pressure on her pinched arteries. May refused. 'After three months of complete rest, I can now function,' she said, 'but I can't even carry a purse.'[16]

Inside postal worker Audrey Z. was diapering her six-month-old baby when she noticed her finger was bleeding. She'd stuck herself with a diaper pin and she hadn't felt it at all. She discovered that she had no sensation in most parts of her hands. Electrodiagnostic studies confirmed that she had advanced carpal tunnel syndrome.[17]

Karen B. tried to ignore the increasing pain from her tendinitis until the swelling and loss of function forced her to the doctor. Now just carrying groceries in from the car can leave her arms swollen for days. She can no longer hold a pen long enough to write a short note, or hold a table knife long enough to make a peanut butter sandwich for her young son. Ordinary chores, such as laundry or dishwashing, have become ordeals. A single parent, she used to be a well-paid electronics assembly-line worker. Her tendinitis makes her doubt that she'll ever be self-supporting again.[18]

'I had no warning,' said religion reporter Joan Breckenridge. 'I was just hit by pain. I was off work for three months at first, went back too soon, and then I was off for seven months. And I was unable to do anything at home, including housework – which I consider to be the sole benefit of RSI.[19]

LET US COUNT THE JOBS

Not all MSIs occur in the paid workforce. Carpal tunnel syndrome, for example, was first defined clinically as a disease of housewives. Athletes have been treated for MSIs for several decades. Further, not all MSIs occur in automated workplaces. Sign language interpreters and massage therapists have a high incidence too. Those caveats aside, generally speaking, the more that workers are treated as adjuncts to machines, the more the MSI rate in any workplace goes up.

'Improperly designed tools, inadequate working space, anatomically unsuitable work practices and poor workstation design are known causes of occupational overuse,' wrote Don Couch, director of the Centre for Occupational Health and Safety at the University of Waterloo, for a 1988 article in *OH&S Canada*. 'Often misdiagnosed as arthritis or muscle strain, occupational overuse went practically unnoticed for decades. But in the last 10 years, reported incidences of the conditions have risen sharply ... Light is now beginning to be shed on the proper treatment and prevention of this very real and very serious occupational disorder.'[20]

Dr Tee L. Guidotti wrote in the *American Family Physician* that 'many authorities in occupational health believe that the rising incidence is the result of an increasing number of jobs that require paced or work-driven execution of a limited number of relatively fine motor movements of the hands and arms, as in keyboarding, assembling small parts, cutting fabric and packaging small items.'[21]

As well as developing in individuals after years of gradual injury, MSI outbreaks sometimes occur suddenly, after the introduction of new equipment, or some other change in the workplace. A Singapore refrigerator factory, for example, had no MSI complaints until one component in the assembly suddenly doubled in size – and workers started crashing.[22] Even without any change in the workplace, when one worker pushes through pain to incapacity, that generally signals that many more workers are already hurting.

Study after scientific ergonomics study has found that workers don't talk much about their aches and pains until somebody asks them direct questions. Employers sometimes try to discredit ergonomics studies with the protest that nobody complained until researchers arrived and started asking workers whether they hurt, and where, and how often.

Some employers claim that personal factors play a major role in susceptibility to MSIs. 'How come nobody integrates into our [risk factor] model what people do when they're not at work?' said attorney Joseph D'Avanzo. 'Two people work side by side. One gets hurts, the other doesn't. Why?'[23] When a worker applies for time off because of an MSI, some employers investigate whether he or she has a systemic disease (such as diabetes or hypothyroidism) or hand-intensive hobbies (such as knitting or rock-climbing). Some employers even pre-test applicants in an attempt to weed out or strengthen new hires who seem to be at particular risk for MSIs.

Still, the main link between jobs and MSIs is statistical. Some heavy jobs are linked to MSIs. The Canadian Auto Workers' union, for instance, has recognized 'overuse injuries' among the hazards that its members might encounter. CAW staff member

George Botic traced the union's interest in MSIs back to 1965, at a General Motors plant in Windsor where 'musculoskeletal injuries were part of the job. Few WCB claims were granted. A few workers got sickness leave.'

In the 1980s, the local Health and Safety Committee managed to hire kinesiologists to try to decrease the rate of injuries. 'There was a big focus on fitness,' Bottic said, 'and it failed miserably.' By 1989, about 35 per cent of the workers at that plant were physically restricted in some way. Botic suggested that a major hurdle in implementing MSI prevention programs was that 'the ergonomics committee was not required to have equal numbers of workers and management representatives. We have to demand joint committees in every large workplace.'

'Most MSK injuries are under- or unreported because they're not recognized. Workers face management intimidation and pressure to produce. They think they can't afford to lose the work time. So, often, the worker avoids hassles and puts up with the pain.'[24]

In 1996 negotiations with the Big Three carmakers, the CAW won important ergonomics protections, including new positions for National Ergonomics Coordinators at Ford, General Motors, and Chrysler.

Some jobs have other hazards so obvious that the MSI risk can be obscured. Meat-packing and poultry-packing jobs involve working in cold warehouses, on fast-paced assembly lines, under strict supervision, with very limited breaks, in the smells and slime associated with killing animals and cutting meat. Workers are often unable to explain just how risky and unpleasant the job is. Turnover is high, lacerations and falls are constant threats. In poultry-packing plants, according to one ergonomist, 'carpal tunnel surgery scars are almost a badge of honour.'

Trucking and transportation clearly involve risks of high-speed highway accidents. Almost 800 U.S. truck drivers died on the highway in 1995. Not so apparent are the stresses on the back (from lifting and shifting cargo) or the demands on the upper body, keeping big rigs on the road even in heavy rains or gale-force winds. 'You've got a fifty-foot kite behind you,' said trucker Janeen Gartner.[25]

Some jobs are heavier work than they appear at first glance. The popular image of nurses, for example, rarely shows them carrying anything heavier than a tray of pills, or perhaps a bedpan. Yet nurses and other health professionals, such as physiotherapists, often have to lift incapacitated patients. 'For years, low-back pain has been considered part of the job,' said Louise Rogers, of the Staff Nurses' Association of Alberta. 'Nursing-specific issues around back injuries seem to be: the frequency with which patients must be lifted; the physical obstructions caused by fragile, extended attachments to patients and mechanical aids; staffing levels, and the pace of work.'[26]

Sewing is another job that involves more lifting and carrying than an outsider might expect. A 1992 article in the journal *Applied Ergonomics* analysed piecework in a trouser factory and found that, in the course of sewing the same short seam 1,500 times a day, 'operators lift an average 406.1 kg of trousers per day and exert an average total force of 2,858.4 kg with the upper limbs and 24,267.9 kg with the lower limbs.'[27]

At another workplace, a tee-shirt factory where Workers' Compensation costs had tripled over a four-year period, strain injuries accounted for 88 per cent of the claim dollars. An insurance company's engineering team helped management and workers identify problem areas, such as non-adjustable workstations. But the main hazard is the seam-stitching position, bent over the sewing machine, gripping the fabric and guiding it under the needle. Worker education helped reduce the carpal tunnel syndrome rate.[28]

Similarly, the heaviest work in a grocery store would seem to be in the back of the store, slinging cases of goods around. Yet *Canadian Grocer*, a trade magazine, reported on a supermarket consortium that discovered its employee injuries were front-end loaded; that is, cashiers suffered more than half the injuries reported by store employees. Among other factors, the health and safety officer found that cashiers lifted, on average, 11,000 pounds (500 kg) of food a day. Other studies have confirmed over and over again that cashiers are at high risk for MSIs. A cashier spends her whole shift standing, reaching, twisting and bending, hoisting bags over counters, lifting parcels and cases, and worst of all, twisting her wrist to scan, scan, scan the Universal Product Codes. Small wonder, then, that *Canadian Grocer* headlined its article on the ergonomics of check-out counters, 'Torture Chamber.'[29]

Still other jobs appear to be light work in the sense that they don't involve actually lifting or carrying materials. As a postal worker, Al MacKinnon said, 'I would code 50,000 keystrokes a day. If you keep multiplying that by the week and month, the number gets quite mind-boggling ... I deal with a lot of people who have been suffering for more than five years, and we can't estimate the cost of what they've been through – the psychological and social pressures on the family, the fact that they can't lift their children or play with them or take part in normal recreational activities ... I often have to convince my members that the pain is not in their head. Many of them have been told this for so many years by specialists, the WCB and their families that they believe it ...'[30]

The highest-paying factory work available to many women who are their families' main breadwinner is electronics assembly work. Twisting and pulling wires and putting parts together doesn't look like heavy work, but the effect of repeating one set of motions all day (sometimes for a twelve-hour shift) – inserting the same little part in the same little socket and securing it – can strain fine muscles as well as eyesight. Similarly, as casinos develop into a major industry, dealers and counters are turning up with MSIs – not because the work is strenuous, but because it's repetitive.

Last but far from least, there's office work. 'The number of workers' compensation claims across Canada for disabling injuries among clerical and related workers has increased approximately 50 per cent since 1983,' pointed out ergonomist Irene Stones and statistician Wendy King in a 1989 *OH&S Canada* article. 'Poorly designed workstations and the fast-paced, boring and repetitive work created by VDTs can give rise to postural problems, eyestrain and stress ...

'... The nagging pain that office workers feel in their arms, wrists, hands, necks, shoulders and/or backs may be the result of a postural problem. Over half of all compensable disabling injury claims among electronic equipment operators are for sprain and strain injuries, which include back pain and other musculoskeletal strains.'[31]

Apart from auto workers, most of the jobs described above are held mainly by women. The high proportion of women affected may well be one reason that WCBs and some doctors are so sceptical that MSIs involve real physiological problems – or if they do, that such problems are related to paid work, and not unpaid activities at home. This scepticism has a chilling effect on people's willingness to report their pain. As the investigation team found at the tee-shirt factory, 'workers were afraid to report injuries ... [because] they would be perceived by fellow employees as malingerers, as crybabies ... '

When injured workers hesitate to seek help or to notify their employers that they're hurt, the problems can multiply. First, the person affected may postpone seeking help until she or he actually crashes, and then faces a long haul to recovery. Second, employers miss their chance to fix problems while they're small. As a doctor's letter in the *British Medical Journal* put it, 'the presence of symptoms in one patient is often a pointer to similar problems in several other people from the same workplace and leads to opportunities for inexpensive modifications that in turn result in reduced sickness absence.'[32]

RISK FACTORS FOR MSIs

Exactly how MSIs develop, and why some people get them while their co-workers don't, is still unclear both medically and ergonomically. Different experts have different opinions.

- Ilene Stones, an ergonomist with the Workplace Health and Safety Agency, firmly stated her view that 'if an activity causes injury, it causes injury to everybody. Some people are more vulnerable than others.'[33]
- Don Couch wrote in *OH&S Canada* about conditions that pro-

mote overuse symptoms, including 'less concrete factors like psychological and emotional stresses. Because of these intangible stresses, overuse symptoms may appear in one worker while another, performing the same tasks, experiences none. This has led to [symptoms] of these disorders being treated as hypochondria, neurosis, and even malingering.'[34]
- In a paper published in the *Scandinavian Journal of Work, Health and Environment*, epidemiologists Barbara Silverstein and Thomas Armstrong et al. postulated a 'cascade model,' weighing how workplace factors interact with personal, social, and cultural factors, as a way to predict which workers are most likely to be hurt.[35]
- Occupational physician Brendan Adams offered a chart of interlocking circles to show how physical, social, personal, and workplace-culture factors combine to injure some workers. Said Adams, 'It's not how often you do a thing, it's how you do it often.'[36] Other experts would say that where you do it – how well the environment and equipment fit – matters too.

JOB RISK FACTORS

In the *American Family Physician*, Dr Tee Guidotti laid out the basics: 'Occupational factors associated with repetitive strain injury include a sustained and awkward posture, excessive manual force, high rates of repetitive movement and unusual or forceful movement of weaker body parts or parts of the body with a mechanical disadvantage in leverage. Load factors have an important role.'[37]

*Repetition*

Some jobs involve making exactly the same movement over and over, such as meat packing or electronic assembly. A quick data-entry clerk may average 10,000 keystrokes an hour, day in and day out. As the general manual from the Workers' Health and Safety Agency Musculo-Skeletal Injuries Prevention Program explains:

# MSIs, RSIs, and the Workplace    19

When the same muscles are used over and over again, they don't have time to recover between repetitions. The blood supply to the muscles decreases, cutting down on the supply of nutrients and allowing waste products to accumulate. As a result, the muscles get tired and start to cramp. If the repetitive motion continues, other muscles start to take over. They, too, may get tired and start to cramp ... The hazards of repetitive work depend on the following factors:

- frequency – how often a repetitive motion must be done;
- speed – how quickly a repetitive motion must be done;
- duration – how long the repetitive work must be done.[38]

*Force, Frequency, Duration*

Force – that is, the amount of effort that the worker applies in every repetition – is another key factor in injuries. A worker who slams staples into paper with the palm of her hand all day (or triggers a staple gun against a wooden surface) is likely to develop an MSI sooner than a worker who clicks away at a keyboard.

Force can also be grouped with two other factors referred to in the Workers' Health and Safety Agency manual – frequency and duration. For the purposes of this book, these three main job risk factors will be abbreviated as the familiar acronym FFD.

*Posture*

Another expert, Barbara Silverstein, gives the key paradigms as frequency, force, and posture or body position. Posture compounds the effects of force and repetition, as well as being a risk factor on its own. Posture may result from workplace design. For example, auto workers who must install parts into a chassis suspended overhead are at risk. So are cashiers who must push parcels along a counter, passing a hand beyond their own body mid-point, or who must lift items to be weighed to scales above their shoulders. Workers who are required to sit still in one position all day, whether in front of a computer or at another machine, may build up what is called 'static loading' strain in their back muscles, which can be injurious.

Posture may also be a matter of personal inclination. People who work with computers often underestimate the importance of how they sit before their machines. 'I used to cross one leg over the other, lean forward, and really crank out the work,' recalled one clerical worker. That was before she crashed. Now she's working on the telephone switchboard, because barring a miracle cure, there's no way that she will ever again be able to type all day.

WORKSTATION COMPONENTS

Other elements of the workplace may put workers at risk, such as a workstation that requires an employee to reach, strain, or lift from an awkward posture. For example, a worker who has to lean across a conveyor belt in order to grab a bundle of materials puts strain on her arms, shoulders, and back.

Tools and equipment may put pressure on susceptible tissues, notably the palm of the dominant hand, leading to nerve compression or entrapment. Or the job may require twisting actions at the wrist, causing the tendons to saw back and forth over the bones, leading to tenderness and swelling. Vibrating tools have been linked to hand and arm disabilities practically since the first jackhammers appeared.

Chilly temperatures have been implicated in injuries, because muscles are more subject to strain when they're cold. Lighting is another crucial element; poor lighting or reflected glare may make workers twist and stoop so that they can see what they're doing.

JOB DESIGN

In addition to the physical set-up of the workplace, job design is a major consideration. At a minimum, people doing repetitive work require frequent breaks – as often as every thirty minutes for intensive work. With computer use, for instance, there are dozens of versions of take-a-break software on the market to encourage workers to stretch or exercise. But some workers refuse to take breaks – pieceworkers, journalists on deadlines – even when they know they should. 'What's better,' said Ilene Stones, 'is to have a variety of duties and positions built into the job.'

When the Canadian Centre for Occupational Health and Safety (CCOHS) reviewed jobs in five widely different workplaces, they found that job design is as important a factor as physical ergonomics. 'In job design,' CCOHS officer Andrew Drewczynski concluded, 'you want to give workers more control over how they do the job.'[39] Varied duties and built-in breaks can be as important in preventing injuries as appropriate workstations and equipment.

PERSONAL FACTORS

Personal factors may also make some people more susceptible to MSIs. Such factors may include chronic diseases, such as diabetes or arthritis, over which workers have little control. One study found a higher rate of MSIs among women workers who had had hysterectomies.[40] Again, any previous injuries to wrists or hands (or to the neck or back, such as whiplash) may increase susceptibility to MSIs. Aches in legs or back may affect the way a person sits or stands, causing extra strain.[41] Avocations may be as problematic as vocations; musicians, dancers, and all kinds of athletes increase their risk for MSIs whether they pursue those activities professionally or for their own pleasure. None of these factors may be described as 'pre-existing' MSIs, in the sense that insurance companies use the term, but they may increase vulnerability.

Since new workers are at higher risk for developing MSIs, some workplaces now perform physical tests on applicants or new hires in hopes of identifying those most likely to crash, and (in the best-case scenario) give them exercises and longer training periods in order to cushion their entry to the worksite. In the worst-case scenario, some applicants have charged that they were not hired because the employer believed their pre-employment tests indicated increased risk for MSIs.[42]

> *Fatigue*
>
> Fatigue may be a factor in MSIs. Tired muscles may be more susceptible to injury. This may be another explanation for the higher rate of MSIs among women workers – most women leave their paid jobs and go home to two or three hours on duty at their unpaid jobs in the home. As the U.N. *Human Development Report 1995* stated, 'Women in both developed and developing countries work longer and harder than men.'[43]

Although pinpointing precisely which activities cause painful conditions in particular workers may be difficult – as some employers insist – any worker with five or more risk factors stands a high probability of becoming an MSI statistic. While the focus on the individual may be appropriate for patient care, it obsfucates public policy issues.

Workplaces could be designed to prevent many, if not most, MSIs. But they are not, for the most part. What is really instructive is to look at how employers and governments respond to MSI epidemics, especially the outbreak that hit Australia in the mid-1980s.

# 2
# Why There Are No RSIs in the Land of Oz (They Go by Other Names)

From the late 1970s to the mid-1980s, some states in Australia were gripped by terror of an MSI epidemic that involved up to a quarter of the Australian civil service at a time.[1]

From 1981 to 1985 at Telecom Australia, the national telephone company, there were 3,976 reports of 'repetitive strain injury' – a rate of 343 per 1,000 customer-service operators, 284 per 1,000 keyboard staff clerks, and 235 per 1,000 'process workers.' Women outnumbered men ten to one in the most affected job categories, and 83 per cent of the RSI reports came from women employees.

Eventually, the Australian government struck a Public Service Task Force and an RSI Committee of the National Occupational Health and Safety Commission to study the problem and recommend solutions. The main recommendation was to remove 'repetitive strain injury' from the list of compensable conditions and to call subsequent cases by various other names, including 'occupational overuse syndrome' and 'regional pain syndrome.'

What actually happened during all the furore? It is hard to find anybody who wants to talk about it now, but various articles have appeared in Australian and international professional journals. Apparently there was a public panic, which caused a widening spiral of worker terror at the threat of disability and management terror at soaring costs. In that panic, methodology went out the window as all parties tried every conceivable technique to alleviate the situation.

REALITY CHECK

Dr Bruce Hocking was the occupational physician for Telecom Australia over a period of fifteen years. His accounts of the RSI epidemic – and his doubts that the term RSI represents any bona fide physical ailment – are widely quoted and echo in many Canadian and U.S. discussions of the issue. In a 1987 article, Hocking discussed ameliorative techniques that were brought into play: 'Various interventions to control the epidemic were tried, including education about posture and the recognition and early reporting of symptoms, training in keyboard skills, the provision of ergonomic furniture, job redesign to include alternate duties, and exercise breaks. From 1983, mixtures of these interventions were instituted at local levels and, around early 1985, on a national basis for telephonists. It is not certain if any or all of these interventions contributed to the large and general decline.' Despite this list of ergonomic changes followed by a drop in incidence, Hocking concluded: 'The contribution of ergonomics to "RSI," at least to type-II "RSI" [e.g., chronic pain] has been overstated within this industry.'

Hocking went on to note a major change in media coverage: 'Education about the symptoms of 'RSI' in 1984 was quite specific in the stages of 'RSI' ... However, during 1985, the emphasis changed to discussion mainly of the workstation, and less attention was given to symptoms, because a view developed that such education was counterproductive and tended to educate staff members into prolonged sick roles as the stages of 'RSI' implied an inevitable progression of the condition ...'

*Description and Suggestion*

Media coverage in the United States and Canada has tended to focus on self-help and ergonomics rather than on descriptions of workers' symptoms. This is a little odd, since the RSI rate at many newspapers averages about 40 per cent of the workforce, and there are plenty of people who are qualified to

> write first-hand accounts. Perhaps North American media are simply trying to avoid the charges made against Australian media, that vivid descriptions frighten people into imagining they have the same symptoms or inspire them to file fraudulent claims for Workers' Compensation. At any rate, although medical journals provide tables listing symptoms, diagnosis and occupations, the reader is unlikely to find anything like that in the mass media – or in this book, for that matter.

Barbara Silverstein commented that there's a lot of conflicting information about what happened in Australia. She noted that 'the medical journals are all full of opinions, not facts.'[2]

OPINION PIECES

Strident opinions abound. In 1986, Dr Leslie G. Cleland published an article titled, 'RSI: A Model of Social Iatrogenesis' in which he argued that 'this syndrome has reached epidemic proportions in a setting where treatments, advice and communications from a variety of professional, official and media sources have created mutually-reinforcing expectations of a high risk of a distressing, disabling condition, which is putatively caused by the use of the upper limbs in tasks that involve repetitive movement or sustained postures. These "treatments" are likely to have a causal role in the development of this syndrome, which may therefore be regarded as an example of "social iatrogenesis."'[3]

This theory – that MSIs are caused by doctors diagnosing them as diseases instead of aches and pains – has been championed in North America by Dr Nortin M. Hadler of the University of North Carolina. His 1990 article, 'Cumulative Trauma Disorders: An Iatrogenic Concept,' did a great deal to frame the whole MSI medical debate in scepticism, long before most family doctors encountered MSIs.[4]

In the same issue of the *Medical Journal of Australia*, Dr Graham D. Wright denounced unions that steered their members to friendly doctors: 'It now stands as a regrettable aspect of the history of occupational health in this country that some unions, who have acted – as they saw it – in the best interests of their members, actively have directed their members with pains that have occurred at work to medical practitioners who would describe and treat these pains as "RSI" ... It is a matter of great regret that some doctors who undoubtedly meant to act in the best interests of their patients have contributed unwittingly to the illness and disability of those whom they meant to help by diagnosing an injury when this did not exist.'[5]

Dr Damian Ireland was even blunter in a 1986 article published in the *Australian Family Physician:* 'RSI is best defined as an occupational neurosis affecting young to middle aged (predominantly female) employees engaged in the low paying, monotonous "low glamour" occupations.' His recommended course of treatment was psychotherapy, but first 'the patient must be prepared to accept that some component of the condition is psychological. An aggressive denial at this stage indicates poor prognosis and renders subsequent treatment ineffective.'[6]

Given the traditional power imbalance between professionals (doctors) and pink- or blue-collar workers, particularly when the doctor is male and the patient is a woman, 'aggressive denial' seems unlikely. Yet apparently Dr Ireland did encounter some women who insisted that their pain was real and was caused by a physical disorder.

PATIENTS' PERSPECTIVE

'Look, I've had five children. I know pain,' one woman told researchers Janice Reid, Christine Ewan, and Eva Lowy, who were attached to various New South Wales community and public health facilities. Their research paper, published in *Social Science and Medicine*, bears an eloquent title: 'Pilgrimage of Pain: The Illness Experiences of Women with Repetition Strain Injury and the Search for Credibility.'[7]

The authors reviewed the social analyses of the RSI epidemic that had already been presented in public health journals and the popular press, but somehow had no influence on medical or official attitudes. Then they synthesized those analyses with in-depth interviews that they conducted with fifty-two women from three industries. They suggest that 'it was in an ideologically polarised environment characterised by doubt, derision and debate that sufferers sought medical advice and treatment ... aspects of the processes of diagnosis, management and referral contributed to the chronicity, unemployment, bewilderment and despair reported by so many.'

The women interviewed did indeed fit Dr Ireland's profile. Most were middle-aged and married, with limited schooling. But their voices, as reported in the paper, hardly fit his stereotype. Rather, their stories convey keen perceptions that they were not believed, much less helped, by most of the medical professionals whom they saw – although there were some doctors, specialists in particular, who believed them and then pronounced them incurably disabled.

Contrary to employers' and some doctors' beliefs that workers were faking RSIs in order to get Workers' Compensation, most of the women in this study delayed seeking help for up to twelve years after they noticed the first symptoms, for fear of peer pressure. They didn't want their 'mates' to call them 'bludgers,' or shirkers. Even when they were sidelined by injuries, some women insisted that all the other RSI claimants were 'bludgers,' but they themselves had simply worked too hard.

Given the confusion surrounding RSIs anyway, and the fact that the workers' injuries were in later stages by the time they sought help, the authors found that 'a large part of their lives became absorbed in seeking advice and care from medical and health professionals who were either as bewildered as they, or in the case of some specialists, frankly sceptical.'

The most common prescription was 'rest,' followed by anti-inflammatories, splints and braces, and surgery. After that came a range of problematic and contradictory prescriptions and suggestions. One woman reported a specialist's comment that it would all go away if she would just get pregnant.

> ### Not an Illness but a Sentence
>
> 'In their search for help and relief the women were caught at the intersection of several influences ... RSI became not an illness but a sentence. Women were judged guilty (and felt guilty) for experiencing pain which could not be located, explained or banished. The attempts to contain the RSI epidemic by discrediting sufferers simply reinforced the fear and hostility surrounding it and inhibited rehabilitation ... [The practitioners'] hostility, as it is refracted through the women's reports, revealed prejudices about semi-skilled workers, women and compensation patients which reflected class and gender conflicts and status differentials in the wider society.'
>
> — Janice Reid, Christine Ewan, and Eva Lowy, 'Pilgrimage of Pain,' *Social Science and Medicine* 32, no. 3 (1990).

## SOCIOLOGICAL ANALYSIS

Sociologist Andrew Hopkins examined the dynamics of the RSI epidemic in a 1990 paper titled 'The Social Recognition of Repetition Strain Injuries: An Australian/American Comparison.' He looked at and discarded the theory that RSIs were the result of epidemic hysteria reinforced by compensation payments. He noted that 'the rise of the problem in Australia coincided with the introduction of new word-processing equipment and the increased work pressures that accompanied this change ... He added, 'the decline of the epidemic has corresponded with the implementation in the workplaces of strategies designed to prevent the occurrence of RSI.'

Hopkins described several factors that served to raise awareness in Australia that were absent or reversed in the United States, including:

- the existence of workers' health centres, established by unions;
- a single, simple, universally recognized term (RSI);

- a generous Workers' Compensation Board (WCB) system that accepted the workers' medical diagnosis without challenge; and
- the fact that the national WCB also gathered and reported all occupational health statistics.

Two major differences in institutional reponses between the two countries are highlighted in Hopkins's report. First, in the United States, the Occupational Health and Safety Administration (OSHA) has been concentrating on blue-collar workers ('for essentially political reasons,' he suggested), whereas the Australian focus was entirely on keyboard operators. Second, since the Australian WCB paid almost any claim a doctor supported, the tendency there was for employers to deny that RSIs were real physiological disorders or caused real disabilities. In the United States, where WCBs look at causality, employers admit that RSIs are real and sometimes permanent disabilities, but deny that they are work-related.

'These comparisons demonstrate yet again the critical nature of the institutional response in generating public awareness and concern,' concluded Hopkins. 'In Australia, a generous compensation system, a regular RSI census, and the existence of the concept of RSI in the popular consciousness facilitated recognition of the problem among keyboard operators. In the U.S., the absence of all these factors has tended to repress any such recognition ... Thus if any policy conclusions are to be drawn ... it is clear ... that the American system is failing and should be liberalized.'[8]

By 1992, when Australian doctors rehashed what had happened earlier, the *Medical Journal of Australia* editorialized that 'many doctors would regard the "RSI epidemic" as a fictitious phenomenon which is best forgotten. The danger with this view is that the baby may be thrown out with the bathwater.' The editorial reminds readers, 'We are left ... with a residue of patients with chronic arm pain related to their work, some of whom are still working reduced hours and some of whom have been off work for months or years and are still in considerable distress.'[9]

# 3
# Waiting Rooms

Given that medical science has not yet reached agreement on exactly how or why MSIs develop, it follows that choice of treatment can only be the doctor's best guess. When MSIs are caught early, at Stage I – when patients seek help for small but persistent pains – then minimal intervention and retraining can have significant results. Entrenched MSIs (Stage III) can be more difficult or even intractable. Without intervention, MSIs often lead to chronic pain, weakness, or loss of function.

> *Hurting at Work?*
>
> Occupational health nurses (OHNs) are the front-line workers in dealing with work-related MSIs. Few workplaces keep doctors on site, but most have nurses available at least part of the time. Since most nurses are women, and women workers are at higher risk for MSIs, workers may be more inclined to tell an occupational health nurse when they hurt than they would be to tell a doctor. Although some workers are suspicious of nurses or doctors paid by their employers, many find the OHN irreplaceable in pinpointing the cause of their problems.
>
> As OHN Pat Bertsche and Dr Thomas Hale explain in the *AAOHN Journal*, ergonomic surveillance is an ongoing activity: 'The health care providers should conduct a workplace walkthrough every month or whenever a particular job task

> changes. This ... allows [them] to: maintain close contact with employees; identify potential light duty jobs; observe individual work practices; and remain knowledgeable about operations described to them by employees.'[1]
>
> Also, a nurse trained to recognize ergonomic hazards can be a liaison between workers and management, identifying problem workstations and suggesting solutions. OHNs find they are spending less time examining bodies and more time measuring the heights of desks, chairs, and workstations.

OUTSIDE THE WORKPLACE

*Potential Barriers to Treatment*

A visit to a primary care physician is an essential first step to finding appropriate care, not only because the doctor in general practice is usually the gatekeeper to specialists, but (more importantly) because the intake doctor should be the person who screens for signs of underlying systemic disorders. Sometimes an aching wrist indicates rheumatoid arthritis, or a complication of diabetes, or hypothyroidism. When the primary care doctor knows the patient well, such possibilities can be verified or ruled out fairly quickly. Otherwise, the first visit should include the doctor's taking an extensive patient history, with questions that might not seem related to upper-limb pain at all.

Unfortunately, questions about workplace conditions rarely arise in the course of a normal examination. 'There is widespread agreement among those involved in occupational and public health that work-related diseases are under-recognized,' according to researcher Michael Lax. He attributed this omission to such factors as lack of professional training, failure to take occupational histories routinely, and reluctance to get involved in Workers' Compensation cases.[2]

For women patients, drawing attention to occupational factors can be doubly difficult. Some patients still report (as the Australian women in chapter 2 did) that they're treated as though their arm pain is psychological in origin. Others say their doctors are simply confused: they first suspect arthritis or other systemic illness, or try to chart whether the pain occurs in relation to the menstrual cycle.

The tendency to blame hormones or lack of them (menopause) for MSIs seems to flow from the fact that two-thirds of MSIs are reported by women. When a condition is specific to women, hormones are usually the first suspect, no matter what other insights the patient may try to suggest. In *Unequal Treatment*, authors Eileen Nechas and Denise Foley argue that women's medical problems are ignored in research, or are studied in men. 'From psychology to oncology, from the emergency room to the family doctor's waiting room,' they write, 'women were being ignored from head to toe.'[3]

Finally, some doctors say outright that they 'don't believe in RSIs.' As occupational physician Brendan Adams put it, 'The occupational literature contains opinions ranging from one extreme (that this is, in fact a fictitious, iatrogenic disorder) to the other (that all occurrences of certain musculoskeletal diagnoses are occupational in cause, and therefore compensable) ... Compounding the controversy, the field is regrettably replete with fuzzy thinking, vague definitions and contradictions.'[4]

*Recognition*

More and more, doctors *are* aware of MSIs, whether or not they're willing to call them work related. Standard treatment is to order a course of non-steroidal anti-inflammatories (NSAIDs), ten to fourteen days' complete rest, and perhaps braces or splints. When the pain seems to centre in the wrist, the doctor may tap the wrist or ask the patient to bend (flex) the joint hard, and wait to see if tingling starts. These are tests for carpal tunnel syndrome – the best known, although not the most common, MSI.

*Incomplete Solutions*

With total rest away from the activities that caused the strain, often mild MSIs show some improvement. However, at the end of the rest period, if the patient returns to exactly the same, unaltered job, the probability is very high that she will be reinjured. (As religion reporter Joan Breckenridge found, even after three months of sick leave her return to work was still premature.) For too many patients, the scenario repeats until they are in such constant pain that their family doctor sends them to a specialist, who probably suggests surgery. Worse, often each specialist consulted delivers a different diagnosis.

The problem here is that medical science does not yet have accurate tools available to identify the specific source of the specific pain reported by a specific patient. The tipoff is the key word, 'syndrome,' as in carpal tunnel syndrome, thoracic outlet syndrome, or tension neck syndrome. When many patients report the same pattern of symptoms and lab results, the medical community can identify the pattern as a 'syndrome,' which translates roughly as, 'Yes, we've seen this before.'

The word 'syndrome' does not connote anything about causation – it does not identify a bacterium or virus, it does not specify the cause of inflammation or discomfort, it does not even link or associate a condition with an activity, as the casual term 'tennis elbow' does. Conversely, the term 'repetitive strain injury' does impute a cause – some would say, wrongly – but does not describe any specific location or pattern of symptoms. Confusion about treatment starts with confusion about diagnosis and causation.

---

*The Five Most Common MSIs*

'The most common RSIs that Workers' Compensation reported, for 1993,' according to kinesiologist Dwayne van Eerd, 'were tendinitis, epicondylitis, bursitis, tenosynovitis and lastly, carpal tunnel syndrome. All these -itises are simply inflammation, and I have to say, that I don't really think that's a diagnosis. Inflammation is really a symptom, not a disease. Carpal tunnel syndrome – that's a symptom too. What's out there now is probably not a lot of accurate diagnosis, not a very good understanding of RSIs, and a very small amount of research.'[5]

SELF-CARE

*Early Stages*

MSI support groups are springing up all around the United States and, to a lesser extent, in Canada, as patients decide to take charge of their own treatment programs. Even with the most knowledgeable doctor or other healthcare practitioner's guidance, the patient's daily activities (including work) will probably be the most important factor in restoring normal function. 'Self-care will make or break your recovery,' is how co-authors Dr Emil Pascarelli and Deborah Quilter put it in their benchmark self-help book, *Repetitive Strain Injury: A Computer User's Guide*.[6]

Based on Quilter's extensive research as well as Pascarelli's twenty years of experience treating musicians with MSIs, their book strongly advises patients to avoid surgery and *not* to wear braces while working, whatever the patient's own healthcare practitioner may order. Rather, the book suggests stretches and exercises to ease stressed muscles and tendons, as a step towards restoring pain-free functionality.

*Ouch! First Aid*

There are a few quick remedies to soothe the flare-up of pain that often catches people off guard, whether it's the first flare-up or only the most recent.

- Ice the affected area. Grab a bag of ice cubes, crushed ice, or frozen vegetables, and rub the bag over the ache for a few minutes. Alternatively, plunge aching hands and wrists into ice-cold water for up to five minutes, or until they're numb. Some people like to alternate cold- and hot-water baths on their arms. Also effective are the generic mentholated gels, marketed under various trade names, that provide a cooling effect.
- Massage the affected area, or have someone else do it. People often massage their aches almost unconsciously. In fact, if you frequently find yourself rubbing your forearms, wrists, or

hands, this may be an early symptom of an oncoming MSI. Better results come from finding a professional massage therapist who can deliver a gentle, soothing massage. Best of all is a deep therapeutic massage that involves at least the neck and shoulders, which are likely to be part of the problem, even when the pain seems localized in hands or wrists.
- Analgesics can help relieve the pain temporarily. Pain relievers that have anti-inflammatory properties (e.g., ASA and ibuprofen) may also help reduce swelling, as sometimes happens with tendinitis. However, analgesics should be taken along with a rest break. Disguising the pain in order to return to hazardous work may lead to more serious injury.

*Reality Checks*

Aching hands and arms, stiff shoulders, necks too sore to twist for a shoulder check – these are wake-up calls. Too many people decline to take them seriously. Yet early response to pain signals can prevent serious injury down the road. On the SOREHAND listserv (an Internet discussion list), people often introduce themselves by listing all their risk factors, explaining that they cannot do anything to cut back on those risks, and asking how they can carry on with their current habits without incurring further injury. It doesn't work that way.

In other settings, people confide that they 'can feel [their] wrists tightening already,' but they don't know what to do to protect themselves. At the very least, early intervention means cutting back on work hours. Self-care in the early stages may include:

- warm-ups and stretches before starting work;
- stretch breaks every thirty minutes, with a walkabout every two hours – for workers who can set their own pace;
- micro-breaks of thirty to sixty seconds, every few minutes, for workers who are under constant supervision;
- daily aerobic exercise, to improve circulation to aching parts;
- exchanging a heavy shoulder bag for a backpack or belt bag and tug-along suitcase;

- trimming long fingernails that may affect hand position on the computer keyboard;
- use of a phone headset rather than cradling a handset on one shoulder, or trying to talk on the phone and type at the same time;
- regular classes in gentle stretching, such as yoga or t'ai chi exercises (videos of these are available for people who don't live where classes are offered);
- attention to nutrition – people who live on caffeine and nicotine don't have as much resilience as people who are reasonably well nourished;
- drinking eight to ten glasses of water a day;
- adequate sleep, to give painful body parts time to heal;
- posture and movement evaluation, and possibly movement retraining in a method such as Feldenkrais or Alexander techniques;
- water therapy, gentle exercise in warm water;
- massage therapy, as a way to improve circulation and also to stretch tight muscles and tendons.

Another very important factor in preventing early MSIs from becoming severe is ergonomic evaluation and adjustment of workstations. Only self-employed people can really call such a program 'self-care.' For most of the workforce, workstation adjustments involve negotiations with supervisors or management, who must in turn acknowledge that a health hazard exists. We'll discuss some of the ramifications of workstation design in chapter 7.

---

*Stop! In the Name of Health*

Wrist position is one factor over which the individual may have some control, especially when entering data into a computer. Ideally, the working wrist should remain in what is called the 'neutral' position: neither bent down towards the palm (flexed) nor back at the wrist (hyperextended), nor outwards (like duck feet), nor inwards (like pigeon toes). Many

> computer operators unconsciously rest their wrists at the base of the keyboard, putting their hands in the STOP! position. STOP is exactly what they should do – what wrist rests are designed to make them do – what braces and splints may be prescribed to make them do, if they don't learn to keep their hands and wrists straight. A negative-tilt keyboard (with the side nearer the computer lower than the side nearer the operator) offers one way to keep wrists up off the desk.

CHARGING OFF IN ALL DIRECTIONS

MSIs are multifactorial – that is, they seem to have a number of different causes that interact together. Therefore, rarely will any one reaction fix the problem. When an outbreak hits a workplace, and people respond quickly to the first signs of MSIs, they tend to try all the remedies they can think of, all at the same time. They fiddle with their workstations, start taking frequent breaks, watch their posture and wrist position, stretch affected parts before they start working, try to get adequate exercise and sleep outside work hours, and perhaps try some alternative therapies such as relaxation techniques, linament, vitamin $B_6$, or herbal remedies. Often, such an energetic response is effective, especially when instituted at the first symptom. Unfortunately, the shotgun approach does little to further medical understanding of just which remedies actually work, for which patients.

MSI support group meetings and online mailing list discussions frequently swap home remedies. Some patients like linament on aching tendons, especially Tiger Balm. Others swear by the Chinese counterirritant known as Salonpas, little plasters that adhere to the base of an aching thumb. There are those who take vitamins, especially B and E vitamins, and those who sleuth for food allergies. Some like fingerless gloves to keep their hands warm.

> *Stretching Back to Health*
>
> MSIs may be described as muscle cramps that went nova. One way to avoid or alleviate cramps is to take frequent

breaks. Another way is to stretch. Ergo, stretching is a suggestion that crops up again and again as a key self-help technique. Two good books with step-by-step instructions are:

- *Conquering Carpal Tunnel Syndrome and Other Repetitive Strain Injuries*, by Sharon J. Butler. Oakland, CA: New Harbinger Publications 1996.
- *Stretching at Your Computer or Desk*, by Bob Anderson, illustrated by Jean Anderson (the couple who created the classic book, *Stretching*). Bolinas, CA: Shelter Publications 1997.

PUSHING THEIR LUCK

Often, MSI patients mean to get medical help, but somehow they keep putting it off – or just hang in at the job while waiting for the specialist's appointment to roll around – until they are in such distress or have experienced such loss of function that the disorder really interferes with their lives and their work. During this interval, the patients are truly 'working wounded.' Some experts now suggest that this delay leads to complications – that by the time a patient arrives in the waiting room, she has not one problem but several.

Unfortunately, a patient's diagnosis may depend more on the speciality of the doctor she consults than on the symptomology she describes. In their paper 'Repetitive Motion Injuries,' published in the *Annual Review of Medicine*, Drs Philip E. Higgs and Susan E. Mackinnon noted that 'the common diagnosis for [cumulative trauma disorder] patients, especially if they are seen by a hand surgeon, is carpal and/or cubital tunnel syndrome. However, the same patients, if examined by a thoracic surgeon, will likely be diagnosed with thoracic outlet syndrome.'[7]

In his book with Deborah Quilter, Dr Emil Pascarelli observed that 'it is not uncommon for people to have five or six different problems at once.' And these problems have generally developed while the patients were actively seeking help. 'Most of my patients saw at least six different specialists before they were properly diagnosed, and they often describe an exasperating odyssey.'[8]

*Find-a-Doc*

Choice of doctor is a crucial element in an MSI patient's quest to return to normal health. When self-help and ergonomic adjustments suffice to relieve symptoms, then a sympathetic family doctor who is willing to write an order for work restrictions may be all the patient needs. However, when surgery is in question, or when an employer disputes a WC claim, or when other non-invasive remedies have failed, then patients start searching for knowledgeable medical care. And such help can be hard to find. Here are some possible shortcuts:

- Workers' Health Centres in Alberta, Ontario, PEI, and Manitoba may be able to offer referrals to local doctors.
- Paul Marxhausen's RSI Page website (http://www.engr.unl.edu/eeshop/rsi.html) includes a Find-a-Doc page (http://www.engr.unl.edu/eeshop/findadoc.html); or by e-mail (mpaul@engrs.unl.edu, with the subject line 'Findadoc'). The website allows RSI patients to post recommendations for U.S. and Canadian healthcare practitioners who have demonstrated expertise working with their conditions.
- The Association of Occupational and Environmental Clinics includes fifty-five member clinics in the United States and Canada of which fifteen or twenty offer specialists in MSIs, including the University of Alberta Occupational Medicine Consultation Clinic with Dr Tee Guidotti in Edmonton and the MFL Occupational Health Centre, Inc in Winnipeg. AOEC headquarters are at 1010 Vermont Ave NW, Suite 513, Washington, DC 20005, and the AOEC website may be found at http://152.3.65.120/oem/oaec.htm.
- Occupational Health Clinics for Ontario Workers, in Toronto, Sudbury, Hamilton, and Windsor, specialize in training family doctors to treat occupational health problems, including MSIs.

> - The Centre for Injury and Disease Response in Toronto (984 Bay Street, Suite 704; tel: 416 944-9008) specializes in MSIs. There is also a Work and Health Research Institute affiliated with the University of Toronto.

SLEEP DISTURBANCES

Disrupted sleep may jolt a person who has been ignoring pain (or masking it) into taking MSI symptoms seriously. Somewhere in the continuum from Stage I (pain after work) to Stage III (constant pain), there's a point where most patients start waking up at night with aches, or tingling hands or arms. A doctor may prescribe wrist or elbow braces at night to prevent the patient from bending or twisting the affected part. Given that fatigue may be a factor in MSIs, then obviously, losing sleep is a double worry. However, the patient may not connect night-time pain with work activities.

Sleep disturbances usually signal a compression of one or more nerves. With all the publicity about carpal tunnel syndrome, doctors and patients tend to suspect the median nerve in the wrist, but often the ulnar nerve is involved instead or in addition. Diagnosing and fixing nerve compression can be tricky – and unless the patient is determined to avoid surgery, suspicion of nerve compression often leads right to the operating room, even when there's no clear diagnosis.

As Drs Higgs and Mackinnon commented, 'the most perplexing disorders are those associated with compressive neuropathies and nonspecific pain complaints. These cases have sparked the most contentious controversy and spawned the most vocal critics of the concept that work in the modern world may produce significant symptomatology. This area is still in need of accepted, effective diagnostic and treatment methods and a clear unifying hypothesis.'[9]

RANGE OF MOTION

The main alternative to surgery – one that is available to mainstream MDs – is physical therapy. Some doctors never suggest it on their own, but they may respond positively when a patient suggests this approach. The standard course of treatment is to try rest, immobilization, and NSAIDs, and then if that doesn't work, to refer to a specialist, who is usually a surgeon. Some health insurance packages in the United States don't even cover physiotherapy (PT).

Physiotherapist Kathy Kilbourn talked about the therapist's role in both evaluating and treating MSIs. 'PT is available only by referral from a doctor,' she said (the requirement has been waived in Alberta), 'and the patient often arrives without a precise diagnosis. We see the patient three times a week and have more chances to identify the basic problem.' She described a process of history taking, observation, manipulation, and testing range of motion, narrowing down the possibilities with every question and every manipulation until she has a tentative diagnosis and a proposed course of treatment. The problem with PT, as she acknowledged, is that 'treatments are limited by what health insurance will pay for.'[10]

In Canada, as well as in the United States, a patient's access to PT may depend on private insurance coverage. Universal healthcare still allows patients unlimited visits to primary-care doctors and (on referral) to specialists, but access to prepaid PT may not be covered. In Alberta, physiotherapy clinics can now accept patients without a physician's referral, but the province pays for only a limited number of patients at each clinic – so physiotherapists advise their clients to check their private insurance policies (if any) or else resign themselves to a waiting list. Given that specialists also have waiting lists, this means that for some patients, the choice of surgery or PT will be determined by which waiting list clears first, rather than by which treatment is more suitable.

PT can be effective even in severe cases. Physiotherapist Peter Edgelow described an exhaustive list of potential problems in an article published in *Physical Therapy of the Shoulder*. He warned that 'there is no quick fix for severe neurovascular entrapment ... The first step in resolution can take 3 months before enough stability and positive results have been obtained for the patient to feel in control.'[11]

Another caveat about PT is that the physiotherapists with the most experience in MSIs tend to be sports medicine specialists. As mentioned in chapter 1, until the last two decades, most MSI diagnoses have been found in athletes and performing artists (such as dancers). However, several sources have warned that sports PT is inappropriate for work-related MSIs. Kinesiologist Dwayne van Eerd used a sports analogy: 'Let's say that a tennis net is sagging because there's a bully sitting on it. Sports medicine would try to lift the net by strengthening the posts. But that won't work, as long as there's a bully sitting on the net.' Started too soon, strengthening exercises can do more harm than good with work-related MSIs.[12]

Occupational therapists may be called in to custom-make braces or splints for an MSI patient, or to analyse work patterns and help devise new ways to work without pain. Occupational therapists (OTs) help patients adjust external factors to accommodate their limitations, whether temporary or permanent.

FIXING THAT PAIN IN THE NECK

Some patients choose chiropractors as their primary healthcare practitioners, or consult chiropractors whether their MDs approve or not. As with any other service, the main question to ask is whether the chiropractor has specific experience in helping patients with MSIs.

'If you decide to seek chiropractic care, seek a chiropractor who has previous experience with the treatment of musicians and/or athletes,' Dr Timothy Jameson, DC, advised in an article for *GuitarBase* magazine. 'Be sure to inquire if the doctor has on-site rehabilitation services and on-site physiotherapy services.' Jameson's clinic in Castro Valley, California, offers a range of diagnostic and therapeutic services. 'Treatment of RSIs should be multifaceted,' he explained. Jameson's book, *Repetitive Strain Injuries: The Complete Guide to Alternative Treatments and Prevention*, was published by Keating Publishers in April 1998.

Here is Jameson's approach to carpal tunnel syndrome:

It is my opinion that 80 to 90 percent of 'carpal tunnel syndrome' cases are not a direct result of median nerve entrapment in the wrist ... my clinical experience has shown that most cases termed 'CTS' are a result of a combination of musculoskeletal and neurological problems stemming from the wrist as well as the arms, shoulders and neck regions. This is why it is necessary to find a competent doctor who will investigate all these areas to determine the extent of injury.

To explain how carpal tunnel syndrome develops, I like to use the following analogy ... imagine the nerves in your arms are like garden hoses, with water (energy) flowing at full pressure when working properly. If you have arthritis in your neck, previous whiplash injuries, misalignments of the spine, or spinal curvature problems, this will cause a 'kink' in the hose, and decrease the flow of water (energy). This will lead to a 40 to 50 percent decrease of nerve flow in the arms and hands ...[13]

AHH YES, RUB RIGHT THERE

Many MSI experts, including Drs Jameson and Pascarelli, recommend massage therapy as a valuable treatment method. Because MSIs are related to muscle tension, and MSI pain sets up a self-perpetuating cycle of tension and reinjury, massage therapy can (by interrupting the cycle) be both soothing and restorative. The primary effect of massage is on blood, lymphatic, and neural circulation, bringing oxygen to tissues and carrying away waste products. Massage therapy can also increase range of motion if it has been limited by injury. Manipulation of trigger points – taut bands in the thin layer of fascia, or tissue over the muscle – has long been used in massage therapy and more recently by some doctors and kinesiologists to treat MSI.

Practitioners have pushed hard over the last decade or so to establish massage therapy as a self-regulating profession – and in many states and most provinces, they have succeeded. British Columbia healthcare insurance even covers limited massage therapy for some specific conditions, such as whiplash. A practitioner who displays a certificate as a registered massage therapist (RMT) or certified massage therapist (CMT) probably has completed a comprehensive course of studies (from 1,000 to 2,500 hours) and will have a thorough understanding of anatomy as well as a soothing touch. While most therapists prefer to work in thirty- or sixty-minute sessions, some now offer on-site massages, which may last only ten minutes, during which the client remains fully clothed and sits forward in a backwards-reclining chair. Massage therapists tend to be insatiably curious about new techniques and constantly add to their repertoires from the ever-expanding field of bodywork.

BODYWORK

The increase in MSIs has stimulated the field of study known as 'bodywork,' a catch-all term that includes more than sixty-five different practices. These techniques usually work directly with muscles, and sometimes skeletal structures, in order to achieve relaxation and improved functioning. Neuromuscular therapy, for example, is massage therapy with an emphasis on a technique called stripping, or narrow compression, to stimulate circulation deep in the muscle grain. Bonnie Prudden's Myotherapy combines massage, trigger-point pressure, and special exercises.

Another hands-on technique is structural integration or Rolfing, named after Ida Rolf, who invented it. A course of treatment involves ten intensive sessions of deep tissue massage and claims to leave clients taller and better aligned than when they started. Treatments can be painful; the technique is not suitable for everyone.

Joseph Heller extended Rolfing to create Hellerwork, which incorporates stretches and exercises to balance the patient's body. Sharon Butler's book on CTS is based on Hellerwork, and Butler regularly trains Hellerworkers who specialize in treating MSIs.

Bodywork also includes ancient techniques such as acupuncture, acupressure, and shiatsu. Asian medicine regards the body as a web without a weaver; when malfunction occurs, the healer tries to find any broken strands and repair them, without necessarily seeking causes.

Yoga and t'ai chi are other ancient practices that promote suppleness and smooth, controlled movement. Deep breathing, which is fundamental to yoga, may be extremely beneficial in itself, since there's some evidence that inadequate circulation contributes to MSIs.

Other kinds of bodywork classes and programs called 'somatics' help people to move more smoothly, gracefully, and economically – which in turn relieves or eliminates the strain that causes MSIs. Among the growing body-in-motion groups are Alexander and Feldenkrais techniques, which concentrate on helping people achieve proper (comfortable) body alignment and movement.

Alexander technique was developed by an Australian stage performer who found that his stance interfered with his breathing and voice projection. Students of Alexander technique practise maintaining a new posture until it becomes second nature. Patrick Stewart (Captain Jean-Luc Picard on *Star Trek: The Next Generation*) is often cited as an ambulatory example of Alexander technique.

Feldenkrais technique was developed by a mathematician and physicist who, laid up with an old rugby injury, began exploring the way his mind and body interacted. He created a set of more than 1,000 exercises (or 'lessons') which, once performed, unconsciously change students' voluntary movements.

Despite the ongoing controversy about whether body work offers long-term remedies, most MSI patients – if they can do the exercises at all – report that they do feel better afterward.

WHO'S AFRAID OF CARPAL TUNNELS?

Your carpal tunnel is a tube about the size of your index finger, which runs from your forearm, between the chunky carpal bones in your wrist, and into the palm of your hand. Inside this tunnel are the tendons that move your fingers, and the median nerve. If the tendons become inflamed, or if there's pressure on the tunnel from (say) water retention or obesity, then the median nerve may become compressed, which irritates it. The ulnar nerve and ulnar artery also pass through the carpals in the same region, but outside the tunnel.

Carpal tunnel syndrome (CTS) has gotten a reputation as the most common MSI. It's not. For awhile it seemed to be the most *provable* MSI, what with the specific tests (Phalen's and Tinel's) and electrodiagnostics, but lately some doubt has been cast on these tests. As well, there are well-established surgical protocols for carpal tunnel release – namely, the surgeon makes an incision and snips the bottom side of the retinaculum ligament that constricts the tendons, thus relieving the pressure against the carpal tunnel. Given the new techniques that have been developed in recent years (e.g., endoscopy), doctors seem to feel that they now know how to treat CTS. Further, patients have come to expect a CTS diagnosis, making it easier for doctors to suggest such interventions.

Especially in the early stages, when intervention is most effective, MSIs can be devilishly difficult to diagnose. Indeed, there's some pressure to acknowledge a new, generalized diagnosis called 'regional pain syndrome.' Some doctors (and patients too) are happier with a diagnosis of 'incipient CTS' or 'mild CTS' than with tendinitis or regional pain syndrome.

---

*SOAP and Slippery Subjects*

Dr Vern Lappi's workshop description of how a physician arrives at a CTS diagnosis illustrates the difficulties in diagnosing tendinitis or other early MSI pain. Dr Lappi explained the SOAP model, which is a mnemonic acronym for:

S – subjective (patient's reported symptoms);
O – objective (signs observed in examination or test results);
A – assessment (seek one explanation that covers all symptoms); and
P – plans (treatment).[14]

---

If MSIs tend to be too slippery for the SOAP diagnotic method, it's not SOAP's fault. MSIs are usually diagnosed mainly from the patient's reports, and these are often confused or confusing. *The single most important thing a patient can do, towards arriving at a clear diagnosis, is to keep a log of symptoms over a period of several days, including location of symptoms, sensations, attempted remedies and effects, and sleep disturbances or new clumsiness or weakness.* MSIs are hard to 'prove' with objective observation or tests. Everybody has arm pain at some point. Barring CTS – or too often, including CTS – diagnosis of MSI is a judgment call.

*Wrists Together for Phalen's Sign*

Dr George S. Phalen's work at the Cleveland Clinic was highly influential in the modern medical understanding of CTS, according to Allard Dembe. From 1947 to 1964, Dr Phalen treated 823 patients with CTS (1,254 hands) – about forty per cent of them surgically – and published a series of articles that convinced most U.S. and European doctors that CTS does have a physical pathology. The downside, Dembe writes, is that Dr Phalen asserted that CTS is 'idiopathic,' without any known cause, and explicitly rejected any link with paid employment or other work. To support his position, his main evidence was simply that the majority of his patients were women.

'Thus,' writes Dembe, 'a seemingly medical and scientific judgment about disease causation was based on a social perspective that regarded males performing work in heavy industry to be the type of patients who could be judged legitimately to engage in strenuous use of the hands. Other work, whether cooking, typing, sewing, or working in an office, was relegated to the domain of 'women's work,' perceived as inherently less demanding upon the hands and wrist.'[15] That lack of perception shades medical understanding to this day.

'In medical school,' said Dr Vern Lappi, 'we were taught that CTS was of unknown origin, typical to overweight middle-aged women.'[16] One of the conservative courses of treatment is to recommend that the patient lose weight – or, if she happens to be pregnant, to stall until the pregnancy comes to term, when the pressure on the carpal tunnel should be relieved. Although a woman may perceive a suggestion that she lose weight as unsympathetic and sexist, in this case, the recommendation actually has some grounding in medical theory.

If CTS is of unknown origin, that tends to support the suggestion that CTS is a symptom, not a disease. In 1992 Dr J. Steven Moore reported on a sweeping review he conducted of CTS research studies since 1860, trying to determine to what extent CTS is work related. Dr Moore concluded that 'the epidemiologic studies reveal a fairly consistent pattern of observations regarding the spectrum and relative frequency of upper extremity morbidity among jobs believed to be hazardous. Carpal tunnel syndrome is but one of these disorders and, in fact, is not the most common one. Disorders of the muscle-tendon unit are far more common, and *carpal tunnel syndrome does not often occur outside the context of co-morbidity*' [emphasis added]. That is, Dr Moore concluded that when CTS occurs, it is usually accompanied by at least one other disorder.[17]

JUST A BUNDLE OF NERVES

Aside from CTS, there's a whole world of other potential MSIs in the upper extremities. Leaving out fascia, blood vessels, tendons, tendon sheaths, and bursae, let's just look at nerves. Web browsers can take a look at the OrthoDoc website, where diagrams illustrate about thirty different muscles that can tense up and produce symptoms that mimic CTS.[18]

Our arms are full of nerves. Like the carpal tunnel, the cubital tunnel is also packed with tendons that can swell and compress a sensitive nerve. Elbows that have to work raised up to reach high desks are very susceptible to epicondylitis – called 'tennis elbow' if it hurts on the lateral side, 'golf elbow' if it hurts on the medial, and sometimes called CTS if the pain travels to the wrist. Either upper quadrant of the body contains a brachial plexus, where muscles, nerves, and blood vessels all jostle together in a tight fit: when inflammation compresses the vascular system or entraps the nerve, pain and impaired motion result. That is usually diagnosed as TOS, thoracic outlet syndrome. Then there is the condition 'double crush,' where one nerve is trapped two different places, proximal and distal (e.g., both closer to and farther from the torso). To complicate matters, there is plenty of evidence of referred pain, that is, pain in the hand–wrist area that originates higher in the arm, or in the neck or back. Just working with palms down for extended periods of time, for example, could cause the pronator teres muscle (at the elbow) to compress the median nerve – and the pain would appear to be at the wrist. 'Think proximal,' urges Dr Adams, meaning try to trace the disorder to where the limb connects to the trunk of the body.

FIRST RESORT, LAST RESORT

CTS surgery won't do a thing to correct any of the conditions described above – but it has been prescribed. MSI patients are often referred to orthopaedic surgeons or to rheumatologists. Rheumatologists tend to look at the hand, up to the elbow. Surgeons tend to recommend surgery. When a person is living with constant pain and a non-functional or only partially functional hand or arm, almost any remedy sounds better than the status quo. One of the most frequent questions to SOREHAND and other Internet MSI discussions goes something like this: 'I've been diagnosed with carpal tunnel syndrome and advised to have surgery to correct it. What should I know about CTS surgery?'

What everybody should know about CTS surgery is that it should be a last resort. There are very few long-term studies on the success of Carpal Tunnel Releases. Although the surgery itself can be done with local anaesthetic on an outpatient basis, recuperation usually involves some discomfort and significant inconvenience. Moreover, if the patient returns to the same working conditions that produced the condition, even surgery won't prevent a recurrence.

Drs Higgs and Mackinnon looked at the results of CTS surgery and concluded, 'In the majority of patients ... carpal tunnel syndrome is responsible for only a small component of symptoms ... The tragic cost of the persistence of other symptoms may be multiple subsequent operative procedures as the surgeon first repeats the carpal tunnel decompression, then proceeds to decompress the median nerve in the forearm and the ulnar nerve at the elbow or even the brachial plexus in the thoracic outlet.'[19]

> *Before Surgery – Share This with Your Doctor*
>
> Dr Susan E. Mackinnon is a plastic surgeon and has been a professor of plastic surgery at the Washington University School of Medicine. She has published extensively on MSIs, sometimes co-authoring reports with physiotherapist Christine Novak. Washington University distributes reprints of her article about the pathogenesis of carpal tunnel syndrome in the September 1994 issue of the *Journal of Hand Surgery*. The package is titled 'Patient Education,' but it offers instructions for doctors too. This article is a must-read for doctors whose entire repertoire for handling MSIs consists of rest, NSAIDs and immobilization, followed by surgery.
>
> An accompanying press release carries this statement from Dr Mackinnon: 'We find that our patients get much better with conservative management ... Based on our results, surgery often is not necessary, nor recommended.' Instead, she concludes, 'The vast majority of problems can be controlled, if not completely eliminated, with a simple but strict exercise program and modifications of positions and postures of the head, neck, back and arms that put nerves at risk for compression and muscles at risk for unhealthy patterns of imbalance.'
>
> Anybody who feels pressured to consent to surgery could benefit by sending for a free copy of 'Patient Education for Cumulative Trauma Disorder,' by Susan E. Mackinnon, MD, available from Washington University School of Medicine, Campus Box 8508, 4444 Forest Park Ave, St Louis, MO, USA 63108-2259.[20]

## WHAT'S SO WRONG WITH SURGERY?

Although surgery has been standard treatment for CTS since the 1950s, these days even organizations such as the American Society for Surgery of the Hand are recommending conservative treatment instead. ASSH president Dr Dean Louis told the 1997 Managing Ergonomics conference that 'most cases will respond to nonsurgical methods.'[21] Dr Gary Franklin, Medical Director of Washington State Department of Labor and Industries, took a harder line, and talked about a 'syndrome of the spreading diagnosis.' He said that Washington L&I puts tight restrictions on medical procedures for MSI clients applying for Workers' Compensation, because otherwise, 'the surgeons march up and down the patients' arm and leave them severely disabled.'[22]

Actually, carpal tunnel release is a fairly well-established procedure, compared with some of the surgical procedures that have been recommended to MSI patients. Any surgery that touches a nerve can lead to a condition called reflex sympathetic dystrophy (RSD), which one SOREHAND member characterized as 'constant, unremitting, shrieking pain.'

Karen Strauss of New York established the Reflex Sympathetic Dystrophy Network, with a highly organized and informative website at http://www.rsdnet.org. Strauss's fact sheet[23] states that 'there is no cure for RSD,' and that 'only twenty percent of patients are able to fully resume prior activities.' While RSD can be caused by accidents or illness, at least some cases are iatrogenic.

Although it is not clear whether the numbers of RSD cases have risen in recent years, the science of pain management grows ever more definitive. The journal *Pain* reported in October 1995 that a special consensus conference agreed on a new umbrella term, complex regional pain syndrome (CRPS), for a wide range of conditions, including the disorders formerly called RSD and causalgia. 'Two types of CRPS have been recognized,' according to the *Pain* article. 'Type I corresponds to RSD and occurs without a definable nerve lesion, and type II, formerly called causalgia, refers to cases where a definable nerve lesion is present.' Besides observable injury to a nerve, cases are categorized by such findings as abnormal skin colour, temperature change, or edema. Treatments tend to be intensive, including intravenous morphine, anaesthetic or pharmacologic nerve blocks, or surgical severing of the sympathetic nerve.[24]

Pam M. regrets the day she consented to surgery. She posted an account of her experience to the SOREHAND listserv and gave permission for its use here:

I also have ulnar neuropathy in my dominant right arm and hand. I was diagnosed with this after EMG testing in 1988. This injury was from repetitive computer use at my data entry job. My workstation was not adjustable and I had no forearm support. My first symptoms were tingling and numbness in my little and ring fingers and pain in my arm, especially at night. I'm sorry to say I continued keying on the computer for a couple months after my first symptoms while I waited for an appointment with a nerve specialist. The pain got worse in this time.

Pam's first surgery came a year after her initial diagnosis. Unfortunately, her pain only increased. When her doctor suggested a second operation, Pam insisted on a second opinion and drove hours out of her way to get it. The second doctor concurred, so Pam reluctantly agreed. But she knew something was wrong even before her cast came off. For one thing, her fingers bent backwards involuntarily. When she went for electrodiagnostic testing, she got the bad news:

Back home I made an appointment with my rehabilitation specialist who had performed the two previous EMGs. Another EMG was done. I will never forget that day. He examined my arm and hand and seemed concerned. He performed the nerve conduction test and gave me the bad news. The second surgery was unsuccessful, and in fact my nerve was severely damaged – a 95% loss. I was in shock as he held up both hands to compare. My right hand had hollow areas where the muscles had atrophied. I was devastated. I was told nerve regeneration is very slow and with the degree of damage I had, I could not expect improvement.

Over five years have passed since my last surgery. The pain I live with every day now is far worse than the pain I had before either surgery. I get waves of pain so intense in my little finger and ring finger, it takes my breath away. It often feels like there is a fire in my arm. The considerable loss of strength never improved. I have burned my little finger twice because I've lost feeling in it.

If someone is experiencing pain and numbness in their hands I would suggest they not delay in getting medical attention. My biggest regret is that I continued to key thousands of strokes a day even with my symptoms. The 'if onlys' haunt me. I have to accept that my arm and hand will not get better and I must live in pain.

Pam M.'s condition cannot be called typical, but neither is it unique. Long-term studies evaluating the effect of CTS, ulnar, or other nerve tissue surgery are hard to find. Anecdotal evidence indicates that when such surgery goes wrong, the results are catastrophic for the patient. Perhaps patients whose CTS surgery was totally successful simply do not talk about their experiences, or feel any need to join support groups. Still, accounts such as Pam M.'s must be taken as cautionary tales.

MEET THE NEW SPECIALISTS

'The only thing that ever worked for me,' said Calgary occupational physician Dr Brendan Adams, 'is posture and exercise.' And of course, modification of both workstations and job descriptions. Actually, when Adams described his work at an electronics-assembly plant that had a high rate of MSIs among its workers, he talked about a whole spectrum of tools and helpers, in addition to posture and exercise. As coordinating physician, he set up a team to address the worksite problems and enlisted skilled and caring professionals to evaluate and treat affected workers.

Along with an occupational health nurse and a dedicated physical therapist, Adams also lined up consultations with a physiatrist. (That's pronounced FIZZ-i-atrist, an MD who specializes in helping patients overcome body difficulties – also known as PM&R, physical medicine and rehabilitation.) 'A physiatrist works to prevent patients with impairments from becoming patients with disabilities,' explained Dr Ken Hoelscher, a physiatrist from Syracuse, New York. He uses the World Health Organization's definitions: an *impairment* is an impaired body part; a *disability* occurs when an impairment becomes a barrier to self-sufficiency, and a *handicap* is a societal barrier for people with impairments. Physiatrists often work with stroke patients, developing programs to recover functioning, but their expertise is also useful with MSIs.[25]

Also involved with Adams at various times were ergonomists and a kinesiologist. Ergonomists study how to make built environments and job functions friendly to the people who work in them. (See chapter 7 for examples.) Their academic training may start in engineering, public health, environmental health, architecture, or psychology. Psychosocial ergonomists look at job design and workplace social structures. On the biomechanics side, ergonomists generally look at measurements: height, width, reach, adjustability, frequency, load, force. Ergonomists tend to quantify *things* in order to bring them into human scale. The catch is that while degrees are granted in ergonomics, and Board certification is available, at present no licences are required. Anybody may call himself or herself an ergonomist.

Kinesiologists study muscles and humans in motion, and usually have at least an MSc. The speciality comes out of sports and physical education training, but has grown to include a range of studies from worksite evaluation to MSI treatments. When kinesiologist Greg Hart does worksite evaluations, he doesn't measure the workstation, as ergonomists do; he looks at body angles, such as elbow angles. At present, the field is so new that kinesiologists have room to define themselves. (Note: Recently some bodywork practitioners have started using the term kinesiology to describe what they do with clients.)

Not a part of Adams's team, but rising in popularity on this side of the Atlantic, partly as a result of the increase in MSIs, are osteopaths. This is not a new profession, but it's better known in the United Kingdom than in North America. Doctors of osteopathy complete about five years of postgraduate training – more if they choose to specialize – and must pass medical board examinations in order to be licensed. Osteopaths focus attention on the musculoskeletal system, according to the American Osteopathic Association, and offer osteopathic manipulative treatment to 'encourage the body's natural tendency to heal and maintain good health.'[26]

ARTHRITIS BY ANY OTHER NAME ...

When family doctors look at hand–wrist problems, especially in middle-aged or older women, one possibility they have to rule out is arthritis. Indeed, many workers just assume that stiff hands must be caused by arthritis. And in a way, they're right. The medical definition of arthritis is 'an inflammation of a joint,' and the Arthritis Society serves patients with more than 150 different diagnoses. Carpal tunnel syndrome is classified as arthritis. Other kinds of arthritic diseases, such as lupus and scleroderma, are potentially life-threatening.

However, the popular image of arthritis probably involves joint degeneration caused by rheumatoid arthritis or osteoarthritis – which are chronic, progressive, and not considered to be work related. Patients should know that rheumatoid arthritis can usually be confirmed by a blood test, and that osteoarthritis usually shows up on X-rays. If careful history taking does not rule out these diseases, the doctor can double-check with tests.

Another possible overlap is with fibromyalgia, a condition that is also under the umbrella of the Arthritis Society. Severe MSIs often present with trigger points, and the presence of tender points in all four quadrants of the body is one of the defining characteristics of fibromyalgia. Sleep disturbances are another. Fibromyalgia involves constant fatigue and considerable discomfort but is non-progressive and non-degenerative. The National Institute of Arthritis and Musculoskeletal and Skin Diseases is following research into the possibility that fibromyalgia may be linked with abnormally low levels of the hormone cortisol.[27]

As to whether carpal tunnel syndrome and other MSIs should be treated as diseases or as work-related disabilities, at this writing, a spokesperson for the Canadian Arthritis Society said only that the Society is aware of the controversy and recently struck committees to study what role (if any) it should play in future discussions about work-related MSIs.[28]

## CAVEAT EMPTOR

Early in 1996 Pfizer Canada ran an ad in various Canadian business publications. It showed two (women's) hands wringing in pain over a computer keyboard. 'Arthritis is the leading cause of long-term disability in Canada,' said the headline, 'and costs more than $5 billion a year in healthcare services and lost wages.'[29] This was Pfizer's way of announcing its research into new drugs for treatment of rheumatoid arthritis – but any manager whose clerical staff suddenly started wearing wrist braces could easily have gotten the impression that somehow they'd all developed rheumatoid arthritis, which is a personal problem – and not MSIs, which are largely preventable and which some government agencies consider to be management's responsibility.

In a similar vein, TV commercials for certain pain relievers have focused on hand pain: a hat maker in one ad, a guitar player in another, a plumber in a third. All the patients seem to be self-employed. The voice-overs say, 'Our doctor recommended [ibuprofen].' But for the poor meat packer or electronics assembly worker whose hands ache after work, the commercials convey the potentially harmful message that it's all right to mask joint pain in order to continue working long hours. The ad doesn't say, 'See your doctor if your hands hurt.' It says, 'Doctors recommend our product.'

Only a little farther afield are people and companies offering to sell special nutritional guidelines, or herbal or mineral remedies for joint aches, or non-surgical procedures to correct CTS. The proliferation of what might be called 'magic bullets' only underscores the difficulties in treating severe MSIs. In particular, patients should beware remedies that concentrate only on the wrist or on one part of the arm; more and more research indicates most arm problems originate in the shoulders and neck.

Patients will have to make up their own minds about where to spend their money. What must be borne in mind is that MSIs have multifactorial causes, and to date the most successful treatments have been multifaceted, too.

TREATMENT

*I Play Only with My Team*

So many factors are involved with incurring MSIs that a healthcare practitioner would have to be a generalized specialist in order to deal with all the aspects. Actually, what seems to be evolving is a team approach described by Dr Eula Bingham in 1989: 'The ideal diagnostic approach to occupational disease is through a multidisciplinary team approach ... The ideal training ... includes the interaction of physicians with industrial hygienists, epidemiologists, ergonomists, occupational nurses, and health educators ... When this interaction occurs, a mutual respect develops, and a team approach makes the diagnoses more credible and able to withstand the scrutiny that such diagnoses often encounter.'[30]

In Canada, the Bingham model is followed by the Occupational Health Clinics for Ontario Workers (OHCOWs), found in Toronto, Sudbury, Windsor, and Hamilton, and at the Occupational Medicine Consultation Clinic run by the University of Alberta in Edmonton, where Dr Tee L. Guidotti is among the doctors in attendance.

OHCOW clinics don't actually treat patients. OHCOW's practitioners diagnose, and then teach the patient's family physician about appropriate treatment. Then OHCOW really swings into action. 'We look at the worker as a window into the workplace,' said John Van Beek, executive director of Toronto's OHCOW, 'so we can take steps toward taking corrective actions before other workers get sick or hurt ... After our preliminary finding, we put together a report that the worker takes back to the H&S committee. Given the quality of the report we provide it's very difficult for the employer to refuse.'[31]

In the United States, the Association of Environmental and Occupational Health Clinics boasts fifty-five member clinics that have all signed a 'patients' bill of rights' promising information and disclosure, confidentiality, and guidance in dealing with

WCBs and the Occupational Safety and Health Administration. The group's directory includes some two dozen clinics (scattered around the country) that list musculoskeletal disorders among their specialities and that also have teams of practitioners on staff.[32]

Teamwork also prevails at the Clinic of Injury and Disease Response (CIDR) in Toronto. Their clinical approach, as described in an article for the *Canadian Journal of Rehabilitation*, 'has focused on 1) a complete and knowledgeable physical assessment, including physical, psychological and surface EMG components; 2) a multidisciplinary treatment approach; 3) a focus on return to improved muscle function; and 4) an emphasis on patient education and responsibility.'[33]

Trigger-point therapy is an essential element in CIDR treatment regimes. Other tools include massage, movement awareness, and ergonomic workstation assessment. '[We insist] that patients become part of the educational and treatment process,' the CIDR team reported. 'We believe that to a large degree the reason patients often improve in our clinic is due to the sense of control they are given over their illness.'

---

*Sharing Information: Self-Help Groups*

Often, the only way to find out about new treatments or coping techniques is to talk with other MSI patients. Self-help groups are springing up across the United States and, to some degree, in Canada. Internet discussions are accessible to anyone, anywhere, who can get use of a computer. Here are some leads to self-help groups.
- The Association for Repetitive Motion Syndromes (ARMS) is an 'international clearinghouse of information about repetitive and upper extremity injuries' anchored by founder and executive director Stephanie Schoenfeldt Barnes. ARMS offers information packages on such topics as ergonomics and thoracic outlet syndrome. For a $20 (U.S.) first-year membership fee, ARMS members receive a quarterly newsletter chockful of information, resources, and an

updated list of local U.S. support groups. Contact ARMS, P.O. Box 471973, Aurora, CO USA 80047-1973.
- In Toronto, the RSI Support Group can be reached through Injured Workers' Consultants, 815 Danforth Avenue, Suite 411, Toronto M4J 1L2, tel. (416) 461-2411. Web: http://iwc@web.net or http://www.injuredworkers.org. Also see the list of Canadian injured workers groups in the Appendix.
- Paul Marxhausen's Computer RSI Page (http://www.engr.unl.edu/eeshop/rsi.html) offers a link to lists of support groups in the United States and internationally. Contact names and numbers for new support groups are welcome there, too.
- The Typing Injury FAQ (http://www.tifaq.com) also lists groups.
- To join a virtual support group by e-mail, you can subscribe to an electronic mailing list (or 'listserv'), which delivers messages sent by any listserv member to all other members. (With SOREHAND's 750 members, for example, that can mean twenty or thirty e-mail messages a day.) To join SOREHAND, send an e-mail message to Listserv@itssrvl.ucsf.edu. Leave the subject line blank. In the body of the text, write Subscribe Sorehand, Firstname Lastname (where Firstname is your own first name and Lastname is your own last name). Leave the rest of the message blank. Or see http://www.ucsf.edu/sorehand/.

*The New Bureaucracy*

Between patient and appropriate treatment lies one more barrier, imposed by cost concerns. That is, treatment may be chosen more on the basis of what the patient can afford than on the basis of what is most effective. Toronto freelance writer Zoe Kessler said that she really wanted trigger-point therapy but couldn't afford it because she was self-employed without private insurance. She went to a chiropractor instead, and to a Reiki bodywork practitioner, and fortunately got good results.[34]

In Canada, universal health care still allows patients unlimited visits to doctors and specialists. Unfortunately, unless they're covered by Workers' Compensation or their employers, many patients still have to pay out of pocket to get access to alternative therapies or rehabilitation clinics.

In the United States, of course, insurance companies and health management organizations (HMOs) are the gatekeepers for access to medical care. Although HMOs aver that their contribution to medical care is to increase efficiency, they also boast about their corporate profits. Whether the patient's expenses are paid through an employer's policy or through Workers' Compensation, HMO patients report apparently arbitrary limits, such as a maximum of six sessions of physiotherapy per time-lost WCB claim. Considering Peter Edgelow's statement that a patient may need three months of work before seeing any improvement, six visits won't accomplish much. Some doctors chafe at the restrictions HMOs place on them, to the extent that they unionize in order to negotiate, or issue public statements such as 'For Patients, Not Profits,' in the *Journal of the American Medical Association*.

SOREHAND discussions often include requests for advice in forcing HMOs to refer patients to doctors with experience and expertise in handling MSIs. HMOs prefer to keep their clients within their own referral lists. Some patients do see advantages in belonging to organizations that provide a comprehensive range of services, including specialists who might be difficult to find otherwise.

There have been some lively SOREHAND discussions about doctors' fees, with a few posters taking the position that the profit motive is absolutely essential to adequate medical care – an approach that is utterly foreign to the Canadian perspective that medical care is a basic human right. Unfortunately, the predominant tone in most of these discussions is that many patients see their HMOs as adversaries rather than allies in the difficult process of treating their MSIs.

At the very least, HMOs create an internal contradiction in for-profit healthcare. There's a limit to what so-called efficiency can achieve when it comes to chronic conditions. Relations with health insurance companies can also be adversarial: some patients report that they're required to undergo independent medical examinations (IMEs) – stressful, tense examinations in which the insurance companies' doctors are openly sceptical of their reported symptoms – in order to continue their medical care.

BASIC RESEARCH

Perhaps someday research into MSIs will pinpoint exactly what cellular or molecular changes are involved in work-related MSIs. Many practitioners are focusing on changes in the myofascia – the thin layer of tissue that overlies muscle (a bit like plastic wrap on top of bread dough) – which seems to shrink or scar or adhere to itself, although explaining the physical or chemical mechanism by which this happens is still controversial. There's also some evidence that the central nervous system is involved.[35]

For the moment, early intervention remains the key to full recovery. Once an MSI really sets in, treatment issues become difficult. A few MSI patients report excellent, compassionate care all the way. Many more report difficulty at every step of their treatment. It's cold comfort to realize that the increase in MSI patients has been so dramatic over the last twenty or thirty years that there's tremendous financial incentive to develop new, more effective therapies.

# 4
# Bigger Than a Breadbox?

As the Displaced Homemakers' Association used to say, you can do almost anything you want with statistics, but you can't do a damn thing without them. Where MSIs are concerned, statistic gathering has been so haphazard that opponents of ergonomic health and safety regulations can argue with straight faces that not very many workers are affected. So let's try to sort out the confusing factors, and get a grasp on the numbers.

One source of confusion is that definitions differ from one jurisdiction to the next, just as diagnoses differ from one doctor to the next. From the early 1980s to the early 1990s, some occupational health and safety (OHS) specialists tried to estimate the incidence by adding up all the relevant categories such as carpal tunnel syndrome, lower-back pain, or bursitis.

For instance, a document from one Alberta OHS research office to another, dated 28 January 1993, provides seven tables of data painstakingly compiled from the Lost Time Claims (LTC) database for 1987–91, 'since there is no unique code for RSI as the source of injury.' The author noted that the 4,454 cases documented probably do not represent all the actual LTCs due to repetitive motion. Some MSI claims could have been filed as onetime overexertion claims.

Women filed half the MSI claims, although overall, only twenty-three per cent of LTCs come from women – which the author noted could be because of the jobs they fill, or perhaps because 'certain procedures with repetitive motions involve equipment or other features which are better designed for men than for women.' Half the LTCs involved inflammation or irritation of joints, tendons, or muscles. A third involved sprains and strains, and nine per cent involved nerve disorders.[1]

By 1997 Alberta Labour had refined data collection sufficiently to be able to report that 'bodily motion was the largest group of lost-time injuries and disease in 1994 and 1995,' up from 13.8 per cent of LTCs to 14.2 per cent. (The figure, 4,257 for 1995, is pretty close to the earlier estimate.) This category includes 'slipping and tripping without falling, repetitive motion or actions, and activities such as running or climbing.' It does not, apparently, include what other jurisdictions call overexertion, which accounts for another ten per cent of LTC claims.

The Canadian Centre for Occupational Health and Safety's first guidebook on repetitive-motion injuries uses the following figures: 'Over 20,000 Ontario workers received compensation in 1987 for new cases of repetitive motion injuries, accounting for about 600,000 days of lost work.'[2] One of the authors, Andrew Drewczynski, estimated in a phone conversation that the numbers are increasing by ten per cent annually. He also mentioned that a staff member at the Ontario Ministry of Labour had tried to compile MSI statistics from Ontario WCB reports, but 'gave up in frustration.'

There is no central OHS authority in Canada, comparable to the Occupational Safety and Health Administration (OSHA) in the United States, which requires regular reporting of work-related injuries and diseases. WCB categories differ by jurisdiction; for that reason, so do the data. A 1995 article in the *Canadian Medical Association Journal* extrapolated from federal data that one in every thirteen Canadian workers has experienced a work-related illness or injury, including MSIs, chemical exposure, and sick-building syndrome, as well as injuries from heavy work. No longer are blue-collar males the only ones at risk, says an occupational health specialist: 'Work-related illness and injuries are just as likely to happen to women or to white-collar professionals.'[3]

For 1994, CTD*News* cites Statistics Canada as its source for an estimate of 100,000 CTD-related cases in Canada. One very rough way to double-check that statistic is to order and compare time-loss work injuries matrixes from StatsCan. From 1982 to 1989, total time-loss injuries climb from 479,558 to a peak of 620,979; then they taper off to 423,184 in 1993. Among males thirty to forty-nine years old, the same pattern prevails: from 154,844 in 1982, the incidence peaks at 215,162 in 1989, and dwindles to 167,062 in 1993. For females thirty to forty-nine years old – the highest-risk group for MSIs – the rate climbs steadily, without hesitation, from 33,365 in 1982 to almost double that, 61,737 in 1993.[4]

In the United States, OSHA has always required workplace reports of MSI injuries on its '200 log' (injury report form), but the reports haven't always gotten attention. The U.S. Bureau of Labor Statistics (BLS) also collects data separately, by surveying 250,000 workplaces every year. The BLS has produced a chart that shows a steep rise over twelve years, and not always a smooth or predictable rise. Reported cases almost doubled in one two-year period, from 72,900 in 1987 to 146,900 in 1989, and then doubled again to 332,000 cases in 1994.[5] In 1992, the BLS instituted new, more specific questions about MSIs and low-back pain. Out of a new total of more than 700,000 cases a year, about 92,000 cases were designated upper-extremity disorders.

'There has been a fourteen-fold increase [in work-related MSIs] from 1972 to 1995,' said Dr Linda Rosenstock, Director of the U.S. National Institute of Occupational Safety and Health (NIOSH), at the 1997 Managing Ergonomics conference. 'And these are systematic underestimates of this problem.' She noted that NIOSH regards the incidence of MSIs as a 'very significant, serious and largely preventable problem.'[6]

Indeed, some prevention efforts do seem to have taken effect. BLS figures showed an overall drop of about seven per cent from 1995 (332,000 MSIs) to 1996 (307,000), and another eight per cent in 1997 (281,000). Why this should have happened is still open to interpretation.

The AFL–CIO background report noted about the 1995–6 drop that 'the greatest decline in these disorders was reported in industries (auto assembly, meat packing and knit underwear mills) where ergonomic hazards have received a great deal of attention by OSHA, employers and unions.'[7] According to data supplied by the BLS, 1996–7 shows a similar pattern: incidence dropped by about twenty-five per cent in meat packing (from 1,206 to 921 per 10,000), where OSHA has implemented ergonomic regulations, and about twenty per cent in auto assembly (from 885 to 710 per 10,000),[8] where unions and the American

Association of Auto Manufacturers have actively pursed ergonomics programs.

The implications of the rising MSI rate were obvious as early as June 1989, when Tom Lantos, the head of the U.S. Congressional Employment and Housing Subcommittee (and a Representative from California) summed up the results of public hearings with the comment that RSIs constitute 'the number one occupational hazard of the nineties.'[9]

Potentially, almost every occupation in the workforce today carries some risk for MSIs. Blue-collar workers who attend assembly lines all day are at high risk – but so are dentists who manipulate tiny tools with tight, fine hand movements. Typesetters and copy editors, secretaries and computer programmers, cashiers and construction workers, welders and waiters – all categories of workers are susceptible. Office workers have suddenly become high-risk candidates for work-related health problems – an unpredictable and unprecedented turn of events. 'Today, 46 million people use computers on the job,' according to lawyer Susan K. Gauvey, of the Venables (Baltimore) law firm, 'compared to 675,000 about fifteen years ago.'[10]

In 1989 experts estimated that MSIs would account for half of occupational diseases by the end of the 1990s. In 1996, MSIs passed the sixty per cent mark of occupational diseases. Some business representatives still tried to downplay the severity of the problem by pointing out that occupational *diseases* are a small portion of all WCB claims, compared with occupational *injuries*, and that MSIs represented a mere 307,000 cases out of the 6.8 million, in total, reported.

Comparative magnitude is a crucial point in the whole discussion: if there are 423,000 time-loss claims across Canada in a year, and most of them are visible injuries, why should any jursidiction bother to develop OH&S policies to prevent invisible disorders that create only three to ten per cent of all claims? Put another way, if there were about 2.25 million time-loss claims in the United States in 1994, why get excited about the 92,000 upper-extremity MSIs?

Tom Leamon of Liberty Mutual Insurance Company made that very point at the Managing Ergonomics Conference: he cited Liberty Mutual figures that MSIs were only about 2 per cent of LTCs in the private industries where Liberty Mutual acts as WC insurer. (This was not the first time that Liberty Mutual has made the same point; two Liberty Mutual employees, George Erich Brogmus and Richard Markois, released a report in May 1996 that emphasized CTDs of the upper extremities make up only 3 per cent of all Workers' Compensation cases in U.S. industry.)[11] In response, Eric Frumin of the Union of Needletrades, Industrial and Textile Employees (UNITE) noted that Liberty Mutual's 3 per cent was the same as the BLS's 3 per cent and that 'if the employers and insurance companies opened up their books and *showed* us what the figures are, we'd all be a lot better off.'[12]

---

*Highest Bid*

CTD*News* reported the results of its own survey in May 1993, providing much higher figures than found elsewhere. 'One in eight American workers have been affected,' according to the headline. And overall, 'the rise in CTDs has by now touched at least 14 million American workers.'[13]

---

What may be more important than the incidence of MSIs is their effect. According to the AFL–CIO's 'Background Report on Repetitive Strain Injuries,' 'in 1996, the Work-Loss Data Institute found that three categories of RSIs (sprains and strains of joints/muscles, sprains and strains of part of the back and carpal tunnel syndrome) ... combined represented more than 47 per cent of all work related disabilities and more than 54 per cent of the total disability days.'[14]

The same source argues that 'RSIs are crippling injuries. They account for 30 per cent of all lost-workday injuires, or more than any other single category ... Repetitive motion injuries lead all events and exposures in median days away from work.' And just in case there was any doubt, the AFL–CIO notes that 'women are ... five times more likely than men to develop carpal tunnel syndrome and almost three times more likely to develop tendinitis.'

Another reason for concern is that MSIs have been increasing at the same time that workplace injuries in general have been declining. The *Globe and Mail* described a 1996 Statistics Canada report that treats MSIs as work-related injuries, not diseases.

StatsCan found that the rate of workplace mortality has dropped (from 918 fatalities in 1970 to 709 fatalies in 1994) and the rate of workplace injury has decreased from 11.3 per cent to 6.9 per cent of workers in the same period.[15] However, since MSIs require longer recovery times than other kinds of injuries, because medical treatment for advanced MSIs tends to be lengthy and somewhat problematic, the net effect is that insurance costs and other indirect costs have climbed.

SO, WHAT'S THAT IN DOLLARS?

'They lose twenty people a week in turnover,' kinesiologist Greg Hart said about one meat-packing plant where he was called in as a consultant. 'That means they lose twelve million dollars a year in training costs. It's the worst kind of work there is, cold, alienating, and repetitious.' It's also injurious, which is why Hart was called in.[16]

'About eighty per cent of my day is spent filling out WCB forms,' said the safety coordinator for another poultry-packing plant, 'and about seventy-five per cent of our lost-time claims are due to RSIs. We lose our experienced workers first.'[17]

Some people will stay with a well-paid job, even after they realize that their co-workers are getting injured. One of the tragedies of MSIs is that they affect people whose personal, economic, or educational situations may not leave them many employment options. But workers who can afford to risk unemployment often bail out. Employers find themselves paying for injured workers and watching uninjured workers move on. Either way, their employment costs rise – sometimes (as we'll see in chapter 8) at a faster rate than if they'd invested in ergonomic consultants and equipment.

Predictably, the earliest mentions of MSIs to be found in business periodicals usually described them as workforce costs, not as medical or health problems. And, perhaps in a bid to catch management's attention, the cost estimates range from astonishing to astronomical. In the United States, the *Financial Post* reported for 1992, lost productivity amounted to seven billion dollars a year.[18] Going one better, the American Association of Occupational Health Nurses cited a 1993 study that put the annual cost at one hundred billion dollars, including indirect costs.[19]

*Business Week*'s account was that 'RSIs in offices and factories are responsible for 56 per cent of all workplace illnesses – 185,000 reported cases in 1990. Workers' compensation claims and other expenses from these injuries may cost employers as much as $20 billion a year, estimates Aetna Life & Casualty.'[20] The *Training and Development Journal* cites an early NIOSH estimate that in 1986 more than five million workers, fully 4 per cent of the workforce, suffered from MSIs. Further, 'the American Academy of Orthopedic Surgeons estimates that resulting lost wages and treatment expenses add up to more than $27 billion annually.'[21]

An article in *Benefits Canada* cited an average Ontario cost of $4,609 per MSI in 1985. By 1989, the number of claims had nearly doubled (to 2,613) and the cost per claim doubled too, to $10,112 apiece.[22] Dr Brendan Adams estimated in mid-1996 that one case of CTS costs the company $50,000 in direct costs and perhaps five times that much in indirect costs.

A caveat is in order: just as with statistics on the actual incidence of work-related MSIs, so too with cost estimates. They vary widely, from source to source and from year to year. Often the definitions are so different from one year to the next that the estimates are difficult to compare.

CTD*News* does an annual survey on MSI costs. In May 1993, CTD*News* estimated that costs had quadrupled since 1987. It noted, for example, that in 1990 average cost of a single case of carpal tunnel syndrome rose to $29,000.'[23]

In June 1996, CTD*News* reported the results of its first annual cost survey: an average of $12,000 per case. 'The findings estimate the direct costs of each CTD case to be $3,720. But when factoring in overtime, employee retraining and production losses due to injured workers, costs more than triple to at least $12,000 per case. U.S. businesses paid about $10.8 billion while the epidemic costs Canadian companies almost $1.5 billion.'[24]

What all the confusion boils down to, in the end, is this clear broth that gels at about one-third to one-half of all WCB expenditures – so far. Figures still can be disputed, just as diagnoses can be. The increase in layoffs, contract workers, and other non-traditional hiring practices increases the probability that some MSIs are not reported.

Who really pays when a worker develops an MSI that could have been prevented? Employers pay all the WCB premiums.

When the worker files a claim and wins WCB approval, the employer may face fines or increased assessments. Those monetary costs may be sufficient to affect the bottom line, to catch management's attention, perhaps even to force a policy change or establishment of a joint management–worker workplace health and safety committee.

There are social costs as well, as Professor Peter Dorman pointed out at the Managing Ergonomics Conference. Dorman analysed potential costs in his book, *Markets and Mortality: Economics, Dangerous Work and the Value of Human Life*. Beyond employers' costs, he found social costs (such as public medical services) and non-monetary costs (such as the loss of the workers' contributions to their families and communities). When employers lobby to cut back on WC premiums and benefits, those costs don't go away; they just get shifted to the public purse.

Indeed, it's the worker who pays the highest cost, in lost income (compensation is lower than the regular paycheque), in lost seniority on the job, in restricted opportunities, and above all, in lost sleep and in pain and suffering and grief. MSIs can and do destroy careers. A resourceful or driven person may pick herself up and find some way to build on her occupational training (minus the use of her hands, of course), but at best the process is lengthy and at worst, it's a setback from which many never recover.

In an age of cutbacks, when governments try to control medical costs by withholding funding, there's no sense in allowing workers to be unnecessarily injured on their jobs to the extent that either healthcare costs rise or other services are cut back. Governments are concerned about healthcare costs – or so they say – yet this is one area where preventive regulations could save millions and billions of healthcare dollars, as well as protect workers.

STOP THE PRESS!

As this book was going to press, Statistics Canada released the results of its second national health survey. Based on interviews with 82,000 people, StatsCan estimated that 'nearly two million people aged 12 and older sustained repetitive strain injuries (RSIs) that were serious enough to hamper their usual activities. These injuries caused by overuse of certain muscles included carpal tunnel syndrome, tennis elbow, other tendinitis and back injury. Injuries to the back or spine accounted for the greatest share (20 percent) of RSIs among men. Injuries of the wrist, hand or fingers were the most common of these injuries (25 percent) among women. Nearly half of all RSIs occurred at work or school.' Two million people would be about 6.6 per cent of Canada's population, or about one in fourteen Canadians. Based on these figures, StatsCan concluded that 'injuries arising from repetitious muscular effort constitute an important health problem.'[25]

Globally, the International Labour Organization counts MSIs among the reasons that 'claims for disability ... are surging in industrialized countries – up to 600 percent in some nations – encouraging governments, private companies and unions to search for ways to get disabled people back to work.' The ILO cited drugs, disease, wars, and malnutrition as primary reasons that an estimated 600 million people now live with mental and physical disabilities. But it also noted that even when nations legislate accommodation for persons with disabilities, some new kinds of impairments have not been helped. 'This is particularly true for workers suffering from the "new" occupational diseases, for instance those related to stress and repetitive injury,' said Ali Taqi, ILO Assistant Director General.[26]

Taqi's words seem to resonate in Canada, where the Canadian Injured Workers Alliance is growing willy-nilly. 'We found about forty injured workers' groups in 1989–90, when we started,' said national coordinator Steve Mantis. 'Now we have eighty-three groups, in ten provinces and one territory.'[27]

As the figures clarify the extent of MSIs, the effects of MSIs still remain murky. MSIs affect us all. We run the risk of MSIs at our jobs. We pay for MSIs through higher prices on the products we buy, through higher taxes to support healthcare for injured workers, and perhaps partial pensions or welfare for those who are permanently disabled.

What is really strange is that society had this whole discussion about OH&S a century ago, and devised a universal program to protect employers and workers alike. The program, still in some sort of operation in most jurisdictions, is called Workers' Compensation.

# 5

# Compensation? But You Don't Look Disabled

BEST SCENARIO

Let's start with an account of cooperative, compassionate, cost-effective management response to the discovery that a worker has developed an MSI.

Donna Lee K. was thirty-three when she discovered she couldn't keyboard anymore.[1] 'I was a Human Resources clerk,' she said. 'Seventy to seventy-five per cent of my time [at work] was spent on the keyboard, maintaining databases, analysing and disseminating information.' In retrospect, she realized that 'I had horrible posture. I sat with my legs crossed and my arms leaning on my legs. I could really crank out the work that way.' When her hands started to hurt, she tried to ignore the pain. She was afraid she would be totally disabled, and eventually she was, but not for the reason she expected: 'It was only after the pain started moving up my arms that I realized it wasn't rheumatoid arthritis.'

Even so, she considered that she was 'very lucky' because her doctor diagnosed her condition right away. 'He was a musician before he became a doctor,' she said. And although she lost use of her hands for the four months between diagnosis and surgery, she's grateful that surgery restored feeling and strength to her dominant (right) hand.

Most of all, she's grateful that her employer, Canada Mortgage and Housing Corporation, 'is very progressive and has wonderful employee policies.' CMHC acknowledged responsibility and maintained Donna Lee's employment throughout her ordeal. 'The Workers' Compensation Board [WCB] sent me to Western Occupational Rehabilitation Centre for three months,' she said, for a retraining program, 'and my company has accommodated me and gotten me new equipment, because I'm permanently disabled now. I still can't open jars now, or turn doorknobs – that turning motion is very difficult. I can't carry groceries. My friend has a child and I can't pick her up.'

Now operating the office telephone centre, Donna Lee is also part of a company committee: 'We're working on implementing [MSI] prevention on a national basis.' She can just barely imagine what her life would be like if she had happened to work for a less progressive employer. 'To have a permanent disability created by your work and then to be thrown to the wolves would be horrible,' she said. 'Who is going to hire a worker who can't use a computer?'

WORST SCENARIO

Shauna C. went through a similar ordeal, but with a much sadder result.[2] 'I've had four surgeries so far, and I refuse to have any more,' she said at the first interview, when she was thirty-three. Not only was she in physical distress, but she was also caught amidst three agencies all trying to duck responsibility for restoring her livelihood. 'The insurance company says I'm fit for my old job, with modifications – the office says that rehabilitation is up to the insurance company – and the WCB says I've used up all my benefits.'

Shauna's office sounds like an ergonomist's nightmare: 'We're a young department, set up with a limited budget and equipment borrowed from other departments,' she said. On a typical workday, she pulls data from computer files while angry or upset people await the information on a phone she holds cradled between her ear and her shoulder. When Shauna first began feeling hand pain, she brought in a doctor's prescription for a phone headset but the department refused on budget grounds. She's a skilled, well-paid worker. 'I get paid $185 a day,' she pointed out, but the company was not willing to spend $200 to protect an employee they valued that highly.

Moreover, costs for WCB far exceed any requested expenses for ergonomic equipment, she said: 'Last year on my worksite there were 600 WCB hours on a staff of eighty-five. I can name twelve people in my office who are wearing arm braces, off the top of my head, and there are four of us who are good and damaged forever.'

At her first interview, Shauna said that her employer was 'trying to fire me,' in order to avoid further expense. Two years later, renewed contact found Shauna was back at her old job, still in pain, still with non-adjustable equipment. But this time, Shauna was a contract worker. After fifteen years in clerical work she was still earning an above-average salary. The catch is, she was no longer eligible for sick leave, benefits such as extra health insurance, or Workers' Compensation. She had no protection at all from being, in Donna Lee's words, 'thrown to the wolves.'

COMPARING CASES

Let's remember that both these women are fairly young, in mid-career, and above all, that before they crashed, they were dedicated and proficient employees. By their own accounts, they fit the profile described in medical literature, which Barbara Silverstein summed up as, 'typically, your high performers.'[3]

Both women have discovered the hard way what too many MSI patients learn too late: when the crash comes, the main factor that determines how you will be treated is not how well you've done your job, nor how severely you're hurt, but rather where you work and what kind of human resources policies your company provides.

Kinesiologist Greg Hart evaluates workstations and helps rehabilitate injured workers. 'I accompany clients to WCB hearings,' he said, 'and it's more difficult than ever to determine who will get Workers' Comp and who won't ... it seems the who is more important than the what.'[4]

## Two Ways to Cut WC Costs

What's happening is that companies (including both Donna Lee's company and Shauna's company) are trying to avoid paying expensive injury claims and WCB fines for unsafe work practices. Donna Lee's company sought to avoid increased premiums or fines by implementing measures intended to prevent future MSIs. Shauna's company took another tack, choosing to deny responsibility. Sad to say, the second method seems to be far more common than the first. But then, that's a modern trend.

### THE ORIGINAL BARGAIN

The Industrial Revolution led to a great social compromise which started in Germany in 1884 and swept across Europe before it arrived on this side of the Atlantic – circa 1910 in New York State, 1909 in Quebec, and 1914 in Ontario. Workmen's Compensation (as it was then called) evolved in order to indemnify employers, recompense workers and their families for work-related deaths or injuries, and rehabilitate injured workers for other jobs whenever possible.

Dennis Guest, a professor in social work, links the rise of WC legislation with the introduction of the Industrial Age: '... the incidence of industrial death, disease, and injury began to climb ... as a result of the application of steam and then, at the turn of the century, of electrical energy to the industrial processes, which required people to work in close proximity to machinery that was heavier, faster, and infinitely more dangerous to life and limb. *These risks were aggravated by a lack of managerial responsibility linked to absentee ownership which characterized the larger industrial enterprises, by the attempt to maximize profits without regard to the welfare of the worker, and by the absence of industrial safety programs.*'[5] That last sentence has a modern echo in transnational ownership, as we shall see.

With the advent of WC legislation, employers promised to pay premiums or carry adequate insurance to compensate those injured or killed on the job. Workers forswore their right to sue for greater compensation through the courts. WCBs were also authorized to impose fines or order changes in unsafe workplaces, and to arrange job retraining for workers who were unable to return to their previous jobs.

## Injured Soldiers of Industry

'The social side of the legislation is to prevent the injured soldiers in the industry and their dependents from being thrown on the scrap heaps as objects of charity,' but this was hardly the main purpose of the act, the Trades and Labour Congress of Canada wrote to Mr Justice William Meredith, Chief Justice of Ontario and Commissioner of Workmen's Compensation, in March 1913.[6]

The Alberta Federation of Labour re-emphasized the Trades and Labour Congress's point in its 1991 publication, *Securing Just Compensation*. '... It is important to recognize that compensation is not a system of social welfare ... It is an insurance system set up by and for employers, and is not some sort of benefit given to workers out of the kindness of the governments' or employers' hearts.

'The Workers' Compensation Act of Alberta, like other compensation acts, was developed because the system of workers suing employers was a costly and unreliable method of securing compensation ... It provides a no-fault insurance system funded by employers as a cost of production in which neither the employer nor the worker assumes blame.'[7]

Employers embraced WC mainly, 'as a scheme which would stabilize costs and spread the risk of liability insurance,' writes Dennis Guest. Also, there was dawning awareness that the workplace itself might be hazardous. Prior to that, 'the conventional wisdom of the day attributed 95 percent of industrial accidents to employee carelessness.'[8]

At first, compensation covered mainly traumas, such as workers who fell from great heights, or were buried in trenches, or were chopped up or electrocuted or trampled, or who suffered any other serious injury. WCB claims forms evolved with the understanding that there was usually one precipitating incident, one drastic accident that incapacitated the worker.

Also, the earliest WC acts covered dangerous occupations, such as factory work – whether the workers were men, women, or children. But, as Dennis Guest observes, other occupations were excluded: farming, wholesale and retail establishments, and domestic service. Note that these are occupations that offer so-called 'light work' deemed more suitable for women.

> *Sickened by the Job*
>
> Justice Meredith addressed occupational diseases in his report: 'It would, in my opinion, be a blot on the act if a workman who suffers from an industrial disease contracted in the course of his employment is not to be entitled to compensation ... By my draft bill, following in this respect the British Act, industrial diseases are put on the same footing as to the right of compensation accidents.' Ontario subsequently recognized six occupational diseases, and so did the first BC *Workmen's Compensation Act* of 1916.[9]

Diseases that develop over time can be difficult to trace back to their origins. The BC board provided two lists, one of diseases and the other of jobs, and when a worker's case could be found on both lists, then there was a presumption that the job had caused it.

Apportioning blame for other diseases, however, challenged the concept of no-fault insurance. Another challenge was that WCBs raised premiums after a certain number of claims, and sometimes exercised their authority to enforce safety regulations by imposing fines. As a result, employers began to dispute workers' claims.

WHO'S IN CHARGE OF HEALTH AND SAFETY?

Originally, enforcing workplace health and safety regulations was up to the local WCB, both in Canada and in the United States. 'In the 1970s and 1980s, this combined role was perceived as a conflict of interest by many provincial governments,' stated the British

Columbia WCB's briefing paper on occupational health and safety (OHS) enforcement. 'The OHS function was transferred to the provincial governments, except in British Columbia and Quebec.'

In the 1990s, that trend has reversed, and in fact a number of separate agencies have re-merged. 'The Yukon Board assumed responsibility for the OHS function in 1992. The function was transferred to the New Brunswick WCB in 1994. Ontario, Prince Edward Island (PEI) and the Northwest Territories assigned the accident prevention or OHS function to their respective workers' compensation boards in 1996, bringing the total to seven jurisdictions – where a WCB administered both functions.'[10]

By contrast, Alberta not only maintains a separate Occupational Health and Safety agency under Alberta Labour, but the government sent up a trial balloon about the possibility of contracting out enforcement to private agencies – a trial balloon that was quickly shot down.

In Ontario, recent changes to the WCB and the *Occupational Health and Safety Act* put more responsibility for illness and accident prevention on the employer and workers – so-called 'internal responsibility' – under the jurisdiction of the WCB.[11] The new agency is called the Workplace Safety and Insurance Board – which, as the Injured Workers' Union has commented, doesn't even include the word 'Compensation.' A top priority for the new WSIB will be 'illness and injury prevention,' according to an October 1997 press release from the Ministry of Labour.[12]

This does not mean ergonomic regulations, or indeed new WCB regulations of any sort. According to a January 1998 Ministry of Labour document, the new Board will pursue its goals through 'a range of motivation and support services,' and the Ministry of Labour will set and enforce standards. Even so, 'the primary and direct responsibility for identifying and addressing workplace hazards, resolving disputes and for complying with the law falls on the owners, employer, supervisors and workers.'[13]

Unions say that 'internal responsibility' hasn't worked in other jurisdictions. The Manitoba Federation of Labour, after a month of public hearings on the Manitoba WCB held across the province, concluded that the 'internal responsibility system' has devolved 'from a supplement to a substitute for enforcement.' Presenting to a Federation hearing, the Canadian Auto Worker's George Botic pointed out that 'in nearly every province, there are more game wardens than health and safety inspectors.'[14]

The Ontario WCB has already incorporated the Workplace Health and Safety Agency (WHSA), which was established as an arm's-length agency by the Liberal government in the 1980s, and which had trained more than 32,000 Ontario workers to recognize MSI hazards in the workplace and respond with ergonomic suggestions.

OSHA STANDARDS

Workers' Compensation agencies in the United States vary from state to state. Some agencies are state-run; some states set WC standards and allow employers to buy insurance from private carriers or to self-insure, and three states (Texas, South Carolina, and New Jersey) do not require WC insurance at all.

Occupational health and safety standards are set nationally. In 1970, according to the AFL–CIO 'Stand Up' campaign, 'President Nixon signed into law the *Occupational Safety and Health Act*, which guarantees every American worker a safe and healthful working environment. Since 1970, the number of people killed on the job has been cut in half and injury rates have fallen by almost 25 percent. But we still have a long way to go.'[15]

In addition to the national standards established by the Occupational Safety and Health Administration (OSHA), states may establish their own OHS agencies, which must have standards at least as stringent as the national agency. State OSHAs have paid serious attention to MSIs since the mid-1980s.

An activist OHS agency, with adequate resources and political support, can wade into troubled workplaces and save both employers and workers some serious long-term grief. In an article in the *American Association of Occupational Health Nurses Journal*, Diane S. Pravitkoff and Joyce A. Simonowitz presented a case study to illustrate the importance of an activist agency – here, the California OSHA.

'Early in 1985 a complaint was filed with Cal/OSHA by an employee of a power supply manufacturing company stating concern about the excessive number of worker injuries attributed to CTD,' according to the authors. Cal/OSHA responded with a surprise inspection of the workplace, starting with a review of the employer's recordkeeping on health and safety reports. This preliminary inspection led to many calls from other concerned employees, reporting specific problems, which guided subsequent inspections.

'In summary,' the authors reported, 'a total of 20 diagnostic entities of CTD were found among 14 workers. Eleven had resulted in surgery, recommended surgery, physical deformity, or protracted pain. Half of the individuals randomly selected for interview had symptoms compatible with CTD.'[16]

Cal/OSHA's medical and ergonomics specialists worked with the employer and workers to re-evaluate workplace, job design, and choice of medical treatments. Investigators found that, up to then, injured workers had been directed to surgery without any effort to ameliorate their conditions by other medical treatments or by modifying their workstations. Moreover, after surgery, they returned to those same unmodified workstations. Cal/OSHA brought in a physician experienced in MSIs to help the injured workers, while an ergonomist recommended process and tool changes, workstation redesign, and rotation of both workers and products.

Some of the fixes that OSHA ordered worked right away; others required some fiddling. Cal/OSHA gave the employer the benefit of the doubt and did not use its authority to fine for noncompliance with work orders. At a subsequent check-up in 1992, Cal/OSHA found that the employer had embraced OHS issues.

As Pravikoff and Simonwitz emphasized, 'The improvements would not have been made without the intervention of the regulatory agency. Once established, the program led to some innovative changes' beyond what Cal/OSHA required.[17]

Oddly, at that time there was not any specific section in Cal/OSHA's mandate that covers MSI hazards or the ergonomically incorrect conditions discovered at this workplace. Rather, the

California Labor Code permits the Chief of OHS to issue special orders in response to particular complaints. Similarly, the federal U.S. OSHA relies on its General Duty clause in order to pursue MSI complaints.

THE CURRENT SITUATION

In the 1990s, WCBs across Canada and the United States have found themselves perilously overextended and under attack – not simply in crisis but, as a 1992 article in *Occupational Health and Safety* magazine stated, 'many crises.' Ever-shrinking benefits and tightened eligibility criteria have led injured workers to acts of desperation, such as the Calgary man who shot himself to death in the WCB parking lot, or the Texas worker who drove his pickup truck into the WCB storefront. Employers complain that premiums cannot keep up with the expense of medical claims and long-term disability payments.

*Boards in Crisis?*

In their article 'The Many Crises in Workers' Compensation,' lawyer Paul Holyoke of the Ontario WCB and Dr Robert Elgie of the Nova Scotia WCB (and former Ontario Minister of Labour) named what they saw as the main problems facing WCBs in Canada and the United States,[18] including:

- service complaints;
- unfunded liabilities of more than fourteen billion dollars in Canada (that is, in long-term obligations to injured workers, beyond what the WCBs have on hand);
- inadequate vocational rehabilitation programs;
- lack of cooperation between unions and management in returning injured workers to new or old jobs;
- adversarial proceedings before WCBs, especially on appeal;
- slumping job markets, caused by recession;
- lack of coordination with other social safety net programs;
- reluctance of physicians to act as gatekeepers to the WCB system.

Holyoke and Elgie fingered one new major expense: '... new claim types that are being accepted by boards across the country ... Take chronic pain for example ... Workers say that WCBs' policies on chronic pain (where they exist) neither recognize the number of workers who suffer nor the extent of their suffering. Employers say that the same WCB policies take into account much more than the work-related pain than they should.'

*What's Straining the System*

'In today's automated workplace, the typical injury is a soft-tissue strain or sprain that may not look serious, but can hurt for months or even years,' wrote journalist Rona Maynard in the *Canadian Business* cover story for February 1993. As an example, she added, 'The Ontario WCB has some 5,500 chronic-pain cases that are costing the system $330 million.'[19]

Note the terminology: in the new debates about WCBs, a phrase that keeps recurring is 'strains and sprains.' That used to mean back complaints, and now it usually includes upper-extremity MSIs. And as we have already seen, pain is not the only issue. MSIs not only 'hurt for months or even years'; they also rob people's hands of their functions.

In another instance, the *Globe and Mail* ran a front-page story on 25 April 1996 headlined, 'Workplace Injuries Drop, But Costs Rise.' The story reported insurance company VP Patricia Jacobsen's analysis that 'more automation and the shift to a service-based economy have meant more strains and sprains, rather than accidents and fatalities on the job.'[20]

The *Globe* cited a study by Statistics Canada that estimated the cost of occupational injuries in Canada at more than ten billion dollars in 1994. (The *AAOHN Journal* cited a comparable figure – one hundred billion dollars annually, or ten times the Canadian amount – for the United States.)[21] Only about half the estimated cost went in direct WCB pay-outs; the rest includes indirect costs. And the *Globe* story suggested that other observers peg the true total of direct and indirect costs at five times the amount cited by StatsCan.

A major sore point for business, said Maynard, is that 'the [WC] board's unfunded liability ... is already $11 billion and growing by about $100 million a month. The board has limped along until recently by jacking up assessment rates with double-digit hikes ... But the reprieve can't last long, because the system is funded solely by levies on employers ...'

The Canadian Labour Congress takes a different view: 'The Ontario WCB has an unfunded liability of about $11 billion,' according to a CLC briefing note. 'It also has about $5 billion in capital reserves. It's a bit like having a mortgage of about $110,000 with about $60,000 in equity built up in the house. You have a house worth $170,000 and a good, steady income which enables you to make the monthly payments. You wouldn't think you had much of a problem, would you?'[22]

Still, employers see WC as costly. And some employers are running scared, or trying to contain WCB assessments by denying responsibility for any worker's injuries, or lobbying the nearest government to roll back WCB regulations so that they exclude MSIs.

THE BIG DENIAL

One way the business community has reacted to the rising rate of MSI claims to the WCB is with increased scepticism about the validity of *all* WCB claims. As upper-limb disorders overtake lower-back pain in the WCB stats, employers seem more inclined to label all invisible disorders as fraud. While MSI patients seek out support groups, searching for cures or at least satisfactory treatments, their employers seek to validate their own perceptions by sharing horror stories about WCB fraud through vivid write-ups in the business and mainstream media.

So pervasive has the scepticism become that politicians have only to say that workers are off on sick leave in order to imply fraud, as Alberta Labour Minister Stockwell Day did in an April 1996 interview with the *Calgary Sun*. Five percent of the province's civil servants were booked off on long-term disability leave, said the minister. 'We're asking the question, "Why are there more people off on long-term disability when workplaces are getting generally more safe and generally more healthy?'" James Forest, head of the Alberta Taxpayers Association, urged the government to scrutinize all long-term disability claims closely. Nowhere in the story is there the slightest concern that the province's worksites might actually be damaging workers.[23]

Since workers also tend to be skeptical when co-workers file WCB claims, this emphasis on fraud encourages people to ignore the early twinges and aches that warn of impending MSIs, and to stay at their jobs until they're seriously impaired.

Indeed, some of the information that employers receive about how to detect WCB fraud sounds remarkably similar to the medical journal articles about detecting MSIs. Consider a fraud-detection program launched in 1995 by the Ontario WCB, which distributed material to employers suggesting they be suspicious of any WCB claim with two or more of the following warning signs:[24]

- first complaint on Monday morning;
- recent employment change;
- healthcare providers and legal services are involved with other claims;
- no witnesses;
- conflicting descriptions;
- history of previous claims;
- claimant refuses recommended medical treatment;
- delayed reporting.

Most of these circumstances arise in work-related MSIs, especially if you accept the theory that MSIs develop progressively. A worker who left work on Friday with some discomfort might wake up Saturday morning unable to move her arm. Also, the Ontario Workplace Health and Safety Agency released a study in May 1995 that showed Monday is the most likely day of the week for sprains to occur.[25] Researchers found that returning from a rest period, such as a weekend or a good night's sleep, leaves workers more susceptible to strains and sprains.

MSIs fit the other indicators too: delayed reporting, conflicting descriptions, and no witnesses might simply mean that the person has already developed several MSI disorders by the time she seeks help. Repeated claims may signify that the worker keeps going back to exactly the same job and getting reinjured.

Finally, the question of refusing recommended medical treatment is most troubling. This so-called indicator of fraud implies that a worker who refuses to undergo surgery must be malingering – despite the fact that many forms of MSI surgery lack documented long-term success.

*Another Perspective on Fraud*

Kinesiologist Greg Hart subscribes to the staging theory of MSIs. 'The original insult [to the body] is never that significant,' he said, 'but then people keep working, and the progression from Stage I is almost inevitable ... The cumulative injuries just build until one day the doors blow off' and the worker crashes.

That said, Hart also has seen workers who present symptoms that do not correspond with any known MSI disorder, or any other disorder, or even with known functions or dysfunctions of the human body. 'I believe in RSI,' he said,

> but there's also a psychogenic component. If you have people complain about a whole arm being numb, that's usually psychogenic.
>
> That kind of description, of general numbness, that's not organically supportable. Unless – it can be psychogenic in another way – say you develop tingling in your ring finger and baby finger and you try to ignore it and you don't seek help. It's not unusual for the mind to create a psychogenic effect that exaggerates the problem so that it makes it impossible for you not to deal with it. The problem is that when you go see a physician and say your whole arm hurts, he won't believe you.
>
> I do know, from my experience working with people, that there are all sorts of people who amplify symptoms for all sorts of reasons, whether it's to convince themselves or if it's malingering tendencies – which I hardly ever see, to be truthful – or it's to draw attention to some other problem at the workplace. Maybe a guy comes to me and says his whole

leg is numb. My first read on this is that for some reason he wants to get out of going to school today. Maybe his supervisor is impossible. Maybe there's some part of his job that can't be done. If you keep sending him back to that job without finding out what his body is trying to tell you, he really is going to get injured sooner or later.'[26]

Dr Barbara Silverstein has a slightly different take on the question of fraud. With a ground-breaking doctorate in MSIs to her credit and already established as an international expert in workplace issues though only in mid-career, Silverstein stated a central truth that few business publications address.

'Most people like being at work,' she said.

They like being useful, they like making a paycheque, they like the social relations they have with their co-workers, and they would prefer to do that. I really don't believe that the vast majority of workers would prefer to be out on Workers' Comp. The reason I say this – it's not just because I have a bleeding heart. Having done lots of physical exams in all kinds of workplaces across the United States, I would find a lot of workers with incredible pain, incredible swelling of tendons, very abnormal conduction of nerves, who want to stay at work. They do not want me to tell their employer that there's anything wrong with them.

*Who's Zooming Whom?*

'There's three kinds of fraud in Workers' Compensation,' said Silverstein. 'There's employer fraud, and that fraud is where employers aren't reporting either the hours worked in the right category or whatever it is that they pay premiums on the basis of. That's the biggest fraud. The second is provider fraud. That's where physicians and other healthcare providers are lying and charging for stuff they didn't do. And the third is worker fraud. It's in that order. And the third, worker fraud, is usually around two per cent. It's worker fraud that gets on TV and everything, but it's really not very common.'[27]

In December 1997, the Sonoma County (California) newspaper, the *Press Democrat*, ran an extraordinarily comprehensive series of articles called 'Insult to Injury,' about Workers' Compensation. In 1993, wrote reporter Mary Fricker, 'business and government leaders, worried about high workers' compensation costs, said up to 300,000 of the claims filed each year, or thirty per cent, were fraudulent. The estimate was widely repeated. But in actuality, in the five years since the state Legislature ordered insurers to set up investigative teams and report all suspected fraud to law enforcement, suspicious claims made up less than one per cent of the total.'[28]

Conversely, in another article, Fricker reported, 'For seven straight years, state auditors found that workers' compensation insurers violated workers' rights in about half the claims it audited.'[29] Similar surveys and audits in other states, such as Ohio and Texas, have turned up similar discrepancies. Fraud by employers, insurers and providers greatly overshadows worker fraud; yet it is the workers who make the news.

*Backlash!*

Then again, there may be another reason that worker fraud gets so much attention – to undercut sympathy for injured workers. For at least a decade, the WCB system has been subject to local and national campaigns to 'reform' its activities, mainly by limiting WCB budgets and mandates. As long-term disability claims have mounted and local WCBs (or insurance companies) have raised assessments to pay out disability pensions, local businesses have responded with demands for budget cutbacks.

'At the same time business associations are trying to block ergonomic protections to prevent RSIs,' charges the AFL–CIO, 'employers and the insurance industry are trying to avoid their workers' compensation responsibilities for the cost of these debilitating conditions.' For example, in 1996 the Virginia Supreme Court upheld a legislative amendment that 'hearing loss and the condition of carpal tunnel syndrome are not occupational diseases but are ordinary diseases of life.' Kentucky excluded from WC any conditions that are a result of the 'natural aging process.'

In Ohio, 'Governor Voinovich signed into law, in April 1997, radical changes in the definition of occupational disease and drastic restrictions in compensation for repetitive strain injuries ... A coalition, led by labor, collected twice as many signatures as required to put the new law to a vote of the people as a referendum issue. Despite massive spending by big business, labor's get-out-the-vote effort turned back this horrible law by a fourteen-point margin (57 percent to 43 percent).'[31]

All legislation is subject to change, of course, and Workers' Compensation regulations have evolved just as the workforce has. Still, the rate of change in the United States has been particularly swift. In 1997 alone, nine states enacted new workers' compensation legislation, according to the American Academy of Orthopaedic Surgeons.[32] Since the mid-1980s, according to Kimberly Patch in the January 1997 *dollars and SENSE* magazine, twenty-eight states have reduced WC benefits, sixteen remained the same, and only six states increased benefits.

*Just for Contrast*

The state of Oregon approached the problem of rising WCB costs with (what might seem to workers to be) an obvious concern. 'State officials recently announced that Oregon's worker injury rates for 1994, measured under OR-OSHA standards, set record lows for both the public and private sectors,' according to a news release from Oregon OSHA.

High on OR-OSHA's list of priorities was preventing MSIs. 'Musculoskeletal disorders (MSDs) have consistently been the most frequent cause of serious injury and illness to employees in Oregon,' states the OR-OSHA newsletter. 'Sprains and strains are in excess of 50 percent of all reported injuries and illnesses ... Often the cost of prevention and control of ergonomic risk factors in the workplace are cost-free or a fraction of the cost of a single claim. In smaller companies, one ergonomically-related claim can mean the difference between being above or below the profit margin.'[33]

BOTTOM LINES

Alas, OSHA doesn't even seem to be in the vocabulary of other WCB-reduction campaigns.

'Help us collect WC horror stories,' urged a lengthy 1996 Internet and faxnet post from a group calling itself the Pennsylvania Chamber Workers' Compensation Network, 'by asking your members to fax/email us a summation of a horrendous WC case with which they have dealt. Governor Ridge wants to see real claims from real companies – we are compiling these and taking them to him.'

Nor is this the first run that the Pennsylvania business community has taken at the WCB. The post observed that 'by targeting medical costs, Act 44 of 1993 stopped skyrocketing increases in workers' compensation costs, but effectively left Pennsylvania with costs that are still too high. Efforts must now focus on the indemnification (wage loss benefits) side of the program. (Medical costs account for about 40 cents of every WC dollar; other benefits account for the remaining 60 cents.)'

The Chamber Workers' Compensation Network cited a Pennsylvania Manufacturers' Association report that 'over $1 billion will be spent this year in litigating workers' compensation cases in Pennsylvania.' And it gave more than 70,000 as the number of disputed WC cases open at the time of writing – more than double the number of open cases in 1990. For some reason, the Chamber Workers' Compensation Network did not even mention the possibility of workplace safety campaigns as an answer to this situation. Rather, it called for 'meaningful Workers' Compensation reform.' Why? 'To ensure that our businesses can *compete* effectively with businesses in other states.'[34]

Meanwhile, just the other side of the state border, the Business Council of New York State had its own Workers' Compensation campaign going. The BCNYS told its members that 'achieving cost savings to make New York employers *competitive* with those elsewhere will require premium reductions of one-third to one-half the existing level ...' (Emphasis added.) Again, this was not a new campaign, according to the BCNYs: 'The business community across New York State has made reducing the cost of workers' compensation a high priority for the past five years.'

There is, of course, another border to the north of New York and Pennsylvania. Sure enough, when the Ontario government introduced its proposed WCB changes in 1995, one of the arguments was that the province's rates were higher than other provinces or neighbouring states. Labour Minister Elizabeth Witmer said it would be 'legislation that is able to deliver fair and generous benefits at a cost *competitive* with other jurisdictions in North America.'[35]

At the bill's second reading, Liberal MPP Richard Patten noted that benefits for injured workers would drop from ninety per cent to eighty-five per cent of net wages, and that even before the legislation passed the provincial parliament, the Ontario government cut employers' premiums by $6 billion.[36]

*An Arm and a Leg*

One of the sorest points for the BCNYs – and business communities elsewhere – is the permanent, partial disability pension. If a worker lost a leg, or most of her or his hearing, that would be permanent (irreversible), partial (not total) disability, presumably for the sort of job that the worker was doing before. When a WCB has obligations to pay out permanent pensions (partial or total) beyond its current assets, the shortfall is called an 'unfunded liability.' The BCNYs takes the position that permanent partial pensions are unfair to employers because 'even workers who could not return to their old job because of an injury [c]ould, over a period of years, be able to undergo training or develop expertise in another kind of job.'[37]

What kind of job? The BCNYs isn't too clear on that goal. But a self-help book by and for injured workers, called *Job Damaged People*, is very clear that a WC claim creates obstacles to returning to paid work. The first obstacle is 'blacklisting, or 'the use of a national system of databases as a tool for screening potential employees. These databases contain names of workers who have been injured on the job, [or] have filed compensation complaints ...'

The next obstacle is, of course, the injury itself, especially for women: 'When a man is injured and can no longer do physical work, there is the possibility that he can still do 'light duty' in an office. However, many women work in a 'light duty' setting to begin with, and the majority of their injuries involve repetitive motion and nerve damage related to office or factory work. If a woman can no longer work in these settings, it is unlikely that she can get a job doing more physical labor. This means that women injured on the job are often knocked out of the labor force.'[38]

One approach to controlling costs associated with MSI claims is to offer the injured worker a lump-sum buy-out instead of a lifelong partial disability pension. In the United States, the offer may be a choice between a lump sum *or* lifetime medical care for the MSI. This latter offer is really tricky, because most U.S. medical insurance policies either bar or charge extra to provide coverage for so-called pre-existing conditions. Most WCBs used to wait two years before they offered such a settlement; now some are pushing for it after only one year.

The New York Committee on Occupational Safety and Health described this trend in a handout about Workers' Compensation settlements: 'Unfortunately, board judges seem to be ... treating RSIs like broken arms that heal with some limitation of movement, instead of like back injuries that represent a continued health risk and potentially severe impairment of activities. The scheduled awards are usually much lower. The additional disadvantage for workers awarded a scheduled loss is that, if aggravation of the RSI puts them out of work in the future, they have to forego wage replacement until the amount awarded to them as a scheduled loss is credited to the carrier and exhausted at the rate of weekly payments – effectively going without pay for as long as it takes. This is consequently a lower percentage of their lost wages, and for people in low wage jobs. More than unfortunately – unacceptably – the trend is to offer these decisions to those with the most to lose.'[39]

---

*Overachievers*

'The reform efforts in the U.S. have tightened the eligibility requirements for WCB in areas such as stress and impairment,' according to *Workers' Compensation Monthly* editor

> Steven Babitsky, 'and that was a cost-cutting measure. In the last few years the reform has reduced the number of benefits, the duration of benefits and number of claims,' all across the board. But the reforms may have gone too far. The January 1996 issue of *WCM* noted that 'in a 30-day period, the states of Colorado, Minnesota, North Carolina and Pennsylvania have all struck down provisions from their 'reformed' Workers' Compensation Acts.' In three of the four cases, the courts ruled that the new legislation was unconstitutional.[40]

Profitability is the key concern of National Council on Compensation Insurance, Inc., a non-profit agency in the United States providing information to private WC insurers. 'For the first time since the early 1980s,' according to the NCCI, 'the U.S. workers compensation system is now financially stable and continues to improve ... After more than a decade of losing money, the bottom line for carriers writing workers compensation insurance is finally approaching the average returns available to investors and companies in other industries ...' Of course, in the United States, the privatized WC system is expected to turn a profit – just as the U.S. health system does.

Comparing the U.S. and Canadian WCBs, the Canadian Labour Congress found that 'Canadian workers' compensation rates are substantially lower than the U.S. rates even though benefits are at least twice as much even in provinces with the lowest benefit levels. WCB assessments are a cheap form of insurance with low administration costs compared with private sector insurance companies.'[41] The Canadian Auto Workers' Union did a similar analysis and resolved that, 'like medicare, we Canadians must defend our better [WCB] system.'[42]

> *Job-Damaged People*
>
> 'Every year, eight to ten percent of the workforce is injured. This means that on the average, workers can expect to be injured three or more times during their careers. Most people

> assume that there is a well-established system that compensates these workers for their injuries. But that is not the case. Most of the compensation systems are full of gray areas, red tape, and loopholes. This situation permits employers and insurance companies to deny, delay or redirect injured workers' benefits. If you were injured on the job, take note – in the world of workers' compensation, nothing much is guaranteed except that you must fight for your rights and the benefits due you by law.'
>
> — From *Job Damaged People: How to Survive and Change the Workers Comp System*.[43]

## CAUTION: FILING A WCB CLAIM CAN BE HAZARDOUS TO SELF-ESTEEM

Typically, a worker has no idea how highly charged the subject is when she first applies for Workers' Comp. Only when a doctor orders a worker to take some time off and rest, and she approaches her employer to validate her claim for compensation, does she discover that she's knocked over a hornet's nest.

Ramonalee K. worked in a poultry-packing plant. When her right arm swelled up, she suspected tendinitis and reported it to her supervisor. 'They put me in a spot where I used my left arm instead,' she said. That sufficed for a couple of days, until her left arm started swelling as well. She left work, went to her doctor, and filed an injury claim with the WCB.

As happens all too frequently, her employer's immediate response to the injury claim was to dispute that the MSI was work related. 'The WCB said I got it from vacuuming,' laughed Ramonalee. 'I don't vacuum. My husband does it.' Then, she said, the company claimed that she'd refused to work, and 'the WCB refused my claim. My doctor was furious.'

In hindsight, Ramonalee realized that she went back to work too soon, to light work, but it still strained her hand. 'I called the union for help,' she said, and got support and information. The second time she filed a WCB claim from the poultry-packing plant, 'it was the same injury but a different job,' she said, 'so the WCB took three weeks to decide if it had to be a separate claim, which it did.'[44]

Kathy Biswanger told the *Calgary Herald* about her three-and-a-half year battle with the Alberta WCB. 'They've spent more money fighting me than they ever will in compensating me,' she said. At one point she had fifty-eight medical appointments in eighteenth months, for tendinitis and what she called radial tunnel syndrome. She believed the MSIs were caused by the thirteen months she worked as a supermarket cashier, scanning groceries, but her former employer disputed the claim.[45]

'Even if you just have a minor injury, get ready to deal with a difficult system,' warned the authors of *Job Damaged People*. 'If your injury or illness is serious, realize that you have just received a new full-time job – to use the system to get well.'[46]

Some workers charge that employers deliberately delay filing reports, or file incorrect reports, when workers file WCB claims. 'There's a pattern of constructive denial,' said Doug Fisher of the Health Sciences Association, a union representing healthcare technicians.[47]

'There's no pre-planned program of slowing down claims,' insisted WCB case manager Ron Pike. Speaking at the February 1993 Edmonton conference on RSIs, he described understaffing and lack of equipment as factors in the delays that both workers and employers find so irksome.

'We have no support staff,' he said, 'so the case manager types letters and cuts cheques, on an ancient computer system that takes three minutes to change a page. There are only twenty-five people to handle sixty per cent of the sixty thousand WCB claims [in Alberta]. And if the employer denies that a worker's injury is work related, then the case manager is duty-bound to investigate. The average wait for a case manager's decision is between two weeks and four months, but some carpal tunnel syndrome cases go up to two years before there's a decision.'[48]

Since that conference, the Alberta government appointed a new WCB chair and gave him a mandate to cut the unfunded liability. In 1996 the government announced that goal was achieved, and gave the chair and board bonuses. In 1997 the chair and board received their third salary increase in three years, bringing the chair's salary to a total of $367,399, nearly triple the WCB chairs of most other provinces.[49] The third-year increase alone was for $92,000 a year.

In fairness, the WCB also raised the non-economic loss (e.g., pain and suffering) one-time lump-sum settlement paid to workers who are deemed to have suffered permanent total disability, meaning they are never expected to hold jobs again, in addition to their monthly payment. 'It went up this year,' said Kelly Eby, WCB media spokesperson, 'from $60,000 to $62,000.'[50]

Meanwhile, said Kevin Flaherty of the Alberta Workers' Health Centre, ordinary workers are finding it harder than ever to file claims or to get benefits. 'The Alberta government says the claims rate has dropped,' he said in a telephone interview, 'but what we're seeing is a lot of direct and indirect coercion. Employers call it "managing claims." Case managers get cash bonuses to keep WC costs down, so they have all kinds of nifty ways to discourage workers who want to make claims.'[51]

APPALLED? APPEAL!

In 1993, Sharon Saunders reported that the Communications Workers of America (CWA) typesetters' local in British Columbia had a high rate of MSIs among its membership and almost every reported case involved a tussle with the province's WCB. 'CWA has yet to get an RSI claim through on first application,' she said. 'We always have to appeal, and we almost always win on appeal.'[52]

A new pattern is emerging with WCB claims: claims seem to be denied routinely, and then granted on appeal. Lawyers are obligatory again, on both sides. Although WCBs were created partly in order to eliminate lawsuits and other delays for the worker applying for compensation, many claimants (especially MSI claimants) find that their employers have retained lawyers to dispute WCB claims and, in consequence, the workers also need lawyers, paralegals, or workers' advocates to assist them in pushing their cases through the WCB system. More and more, the original claim and denial of benefits represent only the first few steps in an intricate dance.

Especially with occupational diseases, the employer tends to dispute claims almost automatically and force the worker to prove that the disorder is work related. In the United States, Kentucky attorney Freeda Steinberg wrote, 'As in all legal proceedings, the injured employee has the legal burden of proving entitlement to benefits under the workers' compensation act ... In general ... if the totality of the evidence demonstrates that the likelihood of work-relatedness is only of equal degree to the likelihood of non-work-relatedness, the claimant will be denied benefits.'[54]

SO, DO YOU KNIT?

When work-relatedness is the question, then the claimant is the person on trial. That may sound backwards, but that's how it works. The claimant's whole life may be under the microscope if the employer is truly determined not to accept any responsibility for the MSI. So if the claim is for epicondylitis, commonly called 'tennis elbow,' then the claimant's fitness activities may be investigated – tennis, golf, squash – and anything that involves swinging the arm may be suggested as a causal factor.

A woman who claims hand-wrist injuries may see her whole crafts or gardening history examined – and sewing, needlepoint, macramé, knitting, or crocheting advanced as the real cause of her pain. Housework, such as vacuuming, ironing, or laundry, not only becomes more painful with an MSI, but may actually be blamed for the injury. Age is often cited: a middle-aged claimant may be described as aging or otherwise infirm. And of course, since so many WCB claims are filed by women, some employers claim that women's hormones are at least as reponsible for workers' MSIs as the kind of work they were doing.

'A number of studies have been performed associating [carpal tunnel syndrome] with various personal factors,' suggested employers' attorney Brooke E. Smith, of Houston. 'The probability that a person will be diagnosed as having CTS increases if the person is (1) female; (2) over the age of forty; (3) overweight; (4) has diabetes, arthritis or hormone disorders; or (5) has ever had a hysterectomy. Personal habits such as use of tobacco, alcohol, and caffeine and failure to exercise regularly also raise the likelihood that a person will develop CTS.'[54]

This emphasis on the personal serves to obscure the institutional factors in MSIs. Millions of women are over forty and are overweight yet don't have MSIs. More millions of people do develop mild aches as a result of recreational or domestic activities, but at least they have the option to *stop* those activities when they feel pain. Generally, workplace demands are what drive people to keep doing work that hurts – until they just can't work any more.

RANGE OF POSSIBILITIES

In Washington state, the WCB guideline is clear: 'If but for the work,' cited Dr Barbara Silverstein, 'the MSI would not have developed, then it's compensable.' This standard shifts the focus away from the personal and back to the workplace, and allows the WCB hearing to consider the tens of millions of other workers who do domestic chores, enjoy sports and hobbies, and even pass through menopause, yet somehow manage not to develop MSIs.

At the other end of the scale, there are jurisdictions where work-relatedness is irrelevant for all the wrong reasons. MSIs are not named as compensable conditions in thirteen states, according to a chart from the Center for Office Technology. And in another twenty-five states, there's only a 'fair' probability of acceptance, or a case-by-case ajudication. In Maine, for instance, 'compensability depends on whether the insurance carrier believes the injury to be work-related or not. If there is lost time involved, the claim must be filed with the state's WC Board. If it is then disputed, the case is adjudicated before the board.'[55]

Although any worker in any jurisdiction whose WCB claim is ultimately denied would be in distress, the implications of such laws are particularly dire for workers in the United States. In that country, odd as it seems to Canadians and Europeans, access to and payment for medical treatment may depend on a successful WCB claim. In Canada, a worker at least has access to prepaid basic medical care, if not necessarily to rehabilitation.

*Canadian Scores*

In Ontario, among other changes, Mike Harris's Proressive Conservative government pledged to 'restore the integrity of the workers' compensation system as a workplace accident insurance plan by precluding compensation for chronic occupational stress and limiting entitlement for chronic pain.'[56] Another important concern for MSI patients is that the Appeals Tribunal will be 'required to apply Board policy in all cases. This change will ensure that the Board has the "final say," as was originally intended.'[57]

Industrial safety and workers' compensation in Quebec are both governed by *La Loi sur les accidents de travail et maladies professionnelles (LATMP)*, which lists specifically which occupational diseases are covered and excludes any others.

'Certain illnesses are presumed work-related if the work includes certain requirements,' according to Quebec lawyer Marie-Anne Roiseux, 'for example, tendinitis, if the work involves repetitive movements. If the injury is presumed work-related, the employer must prove the contrary.' Carpal tunnel syndrome is not presumed to be work related, so the worker must demonstrate that it's work related – or sometimes, the doctor will call it 'tendinitis of the wrist.' There are two levels of appeal.[58]

*Maritimes on Deck*

New Brunswick's Department of Economic Development now woos companies to settle in New Brunswick by offering such enticements as 'the lowest rate of unionization in Canada ... NBTel is the only telephone company in Canada with non-unionized clerical employees ... There has been only one unsuccessful attempt to unionize one centre.' Another inducement involves the availability of part-time workers: 'Because of income and other trends, many families are finding that two incomes are necessary to afford them the quality of life they desire. Permanent, part time employment can fill this need by providing income and allowing flexibility in balancing family and work lives.' (Women constitute two-thirds of all part-time workers in Canada.) Then the kicker: 'New Brunswick call centres take advantage of the lowest legislated fringe benefits in Canada ... A recent modification to the workers' compensation legislation allows virtually all call centre/corporate support operations the rate of $0.25 per $100 of payroll.'[59] One reason for the low rates, according to the Canadian Labour Congress, is that 'changing the definition of entitlement has made RSI claims almost impossible to get in New Brunswick, and in other provinces where the definition of 'accident' has been made more restrictive or where the phrase 'arising out of and in the course of employment' has been changed.'

In Nova Scotia, the WCB commissioned T.J. Murray, Professor of Medicine at Dalhousie University, to research and report on chronic pain. While the author expressed sympathy with patients who suffer chronic pain and urged doctors and the WCB to deal with patients holistically ('It is hard to think of a condition that is not a composite of organic and psychological features'), he apparently had very little experience with upper-extremity MSIs. As a result of Murray's study, the Nova Scotia WCB has implemented a 'functional restoration' program. Briefly, a worker who is still in pain after a deemed 'normal healing time' is offered a four-week course in pain management – and then cut off benefits. The Ontario WCB has expressed interest in implementing a similar program.

*West of All*

In British Columbia, the WCB has always had authority for occupational health and safety. The WCB did draft and circulate proposed ergonomics regulations in 1994 – which met such determined opposition from employers at public meetings that they were shelved and reissued, briefly, just before the 1996 provincial election.

By that time, something else had happened: in July 1995 the provincial government dismissed the entire WCB board of governors, except the president/CEO. On 25 April 1996, it was announced in the Throne Speech that the province would strike a Royal Commission to inquire into the Workers' Compensation system. In consequence, the WCB has been publishing (on the Internet) volumes of concise, informative background papers that review not only BC's policies but policies in other provinces and other jurisdictions as well.

Finally, the provincial government announced a sweeping new Occupational Health and Safety Regulation, effective 15 April 1998, which includes a section on ergonomics requirements. 'The purpose of sections 4.46 to 4.53 is to eliminate, or if that is not practicable, minimize the risk of musculoskeletal injury to workers,' states the new regulation. Not only are the risk factors spelled out, but the onus is clearly put on the employer to assess risk areas and correct potential hazards.[60]

HEARING ROOMS

More and more WCB claimants are finding themselves in courts or hearings, trying to prove their claims. Where MSIs are not eligible for WCB claims, or where workers genuinely believe their employers have done everything they could, or where there are no other recourses available, workers (especially in the United States, but in the United Kingdom too) have turned to civil litigation for redress. Some lawyers and union representatives compare the early litigation on MSIs to early litigation against manufacturers and distributors of other alleged health hazards, such as asbestos and tobacco.

# 6
# So Sue Me

'They've known for years!' charged writer Rory O'Neil in an article for *The Journalist*, the British National Union of Journalists' magazine. O'Neil reviewed the court cases outstanding against computer manufacturers such as IBM, NEC, Honeywell, Wang, Toshiba Compaq, and AT&T, and noted that while no product liability suit had succeeded at the time he wrote, the evidence was already mounting. One very serious allegation to be heard before the courts is that IBM warned its own employees about how to protect themselves against MSIs but did not warn customers.[1]

---

*Computer Warning*

'CAUTION: It is in your best interest to set up a comfortable and healthy workstation. Misuse of your personal computer or failure to establish a safe and comfortable workstation could result in discomfort or serious injury.'

— From *'Creating a Comfortable Work Environment,'* a booklet Compaq provides with each computer purchased.[2]

---

Product liability suits are unusual in Canada. Canadian courts may award compensatory damages to people they find have been wrongly injured, but generally they don't award the kind of whopping punitive damages (often millions of dollars) that the U.S. courts do. Even for lawyers who are prepared to pursue cases in hopes of contingency fees, the amount of preparation required in a product liability suit can be daunting. With class action suits so rare, litigation has to proceed case by case.

That said, Canadian consumers and workers are directly affected by lawsuits in other jurisdictions because they may end up working at equipment that meets the standards of the larger market. Even now, computers sold in Canada come accompanied by the safety advisories imposed by threat of U.S. lawsuits. Or, if by some fluke the United States should implement stricter standards than the Canadian Standards Association, Canadians may end up working with equipment that U.S. makers need to dump.

American workers have used product liability suits as indirect methods to sue their employers. In *Dole* v. *Dow*, for instance, the victim's widow sued an insecticide manufacturer and the manufacturer sued the victim's employer for using its product negligently. When WC legislation restricts workers from direct litigation, yet also restricts eligibility so that the worker's particular injury or illness is excluded, the third-party suit may emerge as a potential remedy. Again, there have been reports from workers whose jobs were downsized away while they were on sick leave; without protective WC regulations, they may consider wrongful dismissal or constructive dismissal suits. In the United Kingdom, at least one worker has successfully sued her employer for negligence and breach of health and safety laws. Muriel Simpson was awarded £186,281 in damages in January 1997.[3]

Whether lawsuits are overused in the United States is a controversial question. Some people – even attorneys – believe that many civil suits are nuisance suits, motivated by lawyers' greed for contingency fees and the hope of an out-of-court settlement. Where MSIs are involved, some say, excessive litigation serves mainly to scare employers away from preventive measures – for fear that any ergonomic renovation or new equipment will be seized upon as an admission of culpability for previous or potential MSIs. Part of the opposition to adopting the (voluntary) American National Standards Institute (ANSI) 365 standard for ergonomic equipment apparently flows from concerns that any such standard would be cited in future lawsuits.

What's clear is that as the rate of MSIs in the workplace rises in the United States, so does the rate of litigation. 'As of late 1992,' lawyer Susan K. Gauvey reported on the Venables law firm website, '... there have been a large number of cases filed against designers, manufacturers, distributors and retailers of computers, keyboards, cash registers, computer 'mouse' devices, stenographic machines, VDTs and supermarket workstations. Over 400 CTD cases are pending in the Federal District Court for the Eastern District of New York alone. Plaintiffs include data entry operators, secretaries, journalists, and supermarket cashiers ...'[4]

As the authority wielded by local WCBs is eroded, the number of lawsuits seems to increase year after year. 'After all,' as the *Wall Street Journal* reported, 'complaints by workers who use computer keyboards or perform other repetitive tasks have jumped tenfold in a decade.' By 1994, the *Journal* could report, 'at least 3,000 high-stakes lawsuits have been filed against equipment makers alleging unsafe design.'[5]

Warnings, such as the booklet *Creating a Comfortable Work Environment* that Compaq now includes with each computer sold, have become commonplace. In a Texas lawsuit, a jury found that Compaq failed to warn users that there were risks inherent in using its products – although the company escaped paying damages when the jury found no conclusive evidence that the plaintiff's carpal tunnel syndrome was caused by keyboarding alone.[6] More recent decisions have revolved around whether the manufacturer supplied warnings and safety information to all customers,[7] and whether an employer removed warnings from workers' equipment.[8]

Susan Gauvey reviewed the barriers to successful product liability suits: lack of precise ergonomics standards for equipment; limited size of claims because MSIs are merely disabling, not fatal; courts' rejection of class action suits; and the difficult of proving a cause-and-effect relationship between using a specific piece of equipment and subsequently developing a specific MSI. She finds that none of these barriers is insuperable. In fact, both official and non-government agencies have been working on ergonomics standards, legislators have been pushing ergonomics regulations, and attorneys have been pursuing all avenues for class actions. 'The medicine and science is still developing,' wrote Gauvey. 'As more is known about CTDs, the risk of liability increases for manufacturers, designers, and sellers of equipment.'

In the summer of 1996 Gauvey's prediction proved accurate when a Maryland court ordered NCR Corp. to pay former supermarket checker Eula Davis $85,000 for medical expenses and lost earnings. Evidence introduced at the trial showed that 'NCR commissioned a videotape showing the correct and safe way to operate the [check-out] scanners and selectively provided safety information to customers. Neither Davis nor her employer received any safety instructions.'[9]

Then on 4 December 1996 the jury in the *Geressy* v. *Digital Equipment Corp.* case awarded three plaintiffs a total of nearly six million dollars, based on Digital's failure to warn users that there was potential for injury.[10] Although two of the three awards were subsequently overturned (one on the basis of new evidence, the other on a technicality), Jeanette Rotolo retained her $274,000 conmpensatory award.[11]

In that case, 'lawyers persuaded jurors that Digital should have warned customers about the dangers of continuous typing – just as it did its own employees,' as legal affairs reporter Reynolds Harding wrote in the *San Francisco Chronicle*. 'The evidence: Company documents detailing how Digital dealt with employee complaints at the same time it was telling the outside world that there is no evidence keyboards can cause injuries.'

Next up, said Harding, would be Apple Computers, Inc., with plaintiffs brandishing the same sort of internal documents as were used in the Digital case. 'It sure makes for a strong case when a plaintiff can rely on internal memos – rather than scientific experts – to show that a company knew the use of its products could cause injuries.'[12]

For U.S. attorneys involved in MSI litigation, Andrews Publications has established a monthly *Repetitive Stress Injury Litigation Reporter*. Here, lawyers (who can afford the subscription price of $685 a year) may find monthly installments in the continuing suits against IBM and Atex, which have sometimes been bundled for class action and sometimes unbundled.

In January 1996 lawyers for plaintiffs Betty Anderson and

Patrisia Gonzales petitioned the court to strike the expert testimony given by a familiar name – Dr Bruce Hocking, of Telecom Australia, whom the defendants had summonsed (see chapter 2). The plaintiffs claim that when he was deposed, Dr Hocking 'explicitly told' them 'that great caution had to be used in trying to extrapolate from Telecom Australia data to anything else.'[13]

In two separate cases, railroad employees lost their suits for compensation for carpal tunnel syndrome. In *Zarecki* v. *Amtrak*, the court found that the plaintiff failed to prove that her work caused her CTS.[14] In *Ciesielski* v. *Norfolk & Western Railway Co.*, the jury accepted the plaintiff's argument that his work caused the CTS, but found that there was no negligence on the employer's part for failing to forestall the injuries.[15]

*U.S. Equal Employment Opportunity Commission* v. *Rockwell International Corp. et al.* offers a glimpse of a chilling future. The EEOC filed a suit on behalf of twelve applicants who claimed that they were rejected for employment at a Rockwell plant, allegedly because pre-employment physical tests showed they either already had or were at higher than average risk for developing RSIs. Rockwell denied that its employment policies screen for disabilities (which would be contrary to the *Americans with Disabilities Act*). Rockwell also argued that, if it were to hire applicants whose medical tests showed impaired nerve conduction, it might be in violation of occupational safety and health regulations. The EEOC responded that the act takes priority.[16] And so the case continues.

In some of these cases, the casual observer has to wonder just what the plaintiffs hope to accomplish. In the last instance, the EEOC's main purpose clearly is not money, but to attack an allegedly discriminatory hiring policy. We may well ask ourselves why individuals and agencies are willing to spend years pursuing these cases through the courts. Maybe they're hoping for a jackpot. Or maybe they just want to stop other people from getting hurt.

# 7
# Staring at the Screen: Computer Workstations

Far from being just sophisticated typewriters, computers have changed virtually every aspect of office work, from spreadsheets to communications to file keeping. They haven't replaced paperwork, of course – on the contrary, computers have multiplied paperwork – but they've made it easier and blindingly faster to produce.

Also, computers have reintroduced occupational health and safety (OHS) concerns to the professional class. Although we know that clerks have suffered writer's cramp since Bernardo Ramizzini described the condition, and that typists were affected in the 1930s (in 1940, Margaret Mettert of the Women's Bureau of the U.S. Department of Labor reported that ailments caused by 'repetitive motion' accounted for one-tenth to one-third of all occupational diseases among women),[1] computers apparently have the dubious distinction of producing MSIs faster than any previous office equipment since the telegraph.

Professionals who work with computers tend to be educated, articulate, socially involved and with reasonably high expectations and self-esteem. When they get hit by MSIs, they are somewhat more likely to seek help or to complain about working conditions than, perhaps, assembly-line workers are. This has led to the perception that computer work carries the highest risk for MSIs. It doesn't. At most, one-third of the MSIs charted by the U.S. Bureau of Labor Statistics are computer related. However, computers seem so innocuous, and white-collar work has seemed so safe, that computer-related injuries have generated a lot of public interest. Who would ever have expected typing to cause permanent impairment?

In fact, intensive computer use has been linked with a whole range of problems, including eyestrain, back strain, miscarriages, and of course MSIs, including carpal tunnel syndrome, thoracic outlet syndrome, ulnar nerve entrapment, and cubital tunnel syndrome. Part of the problem is that the human body was designed to keep moving around, not to sit still for long stretches of time. But the fact is that a lot of computer hardware falls short of being user-friendly. Add to this a workstation originally planned for a typewriter or no typing at all, and perhaps some lighting that glares on the screen – and after a few years the typical worker is liable to encounter aches and pains.

As the Public Service Alliance of Canada manual on computer use states, 'Academic researchers, union experts and equipment manufacturers have all concluded that VDT technology cannot be accommodated by conventional office furniture ... *Components of the work station must be adaptable to the needs of the individual worker in order to prevent fatigue as well as ... numerous health problems* ... Individuals may be able to adapt to poorly designed work situations, but the cost is high in terms of personal physical discomfort and pain.'[2]

---

*Known Risk Factors*

Risk factors with computer use are well documented:

- poor posture, with shoulders slumped and chin poked forward;
- scrunched-up limbs, with shoulders hunched up, knees higher than hips or hands higher than elbows (that is, acute angles at the hip or elbow);
- wrists or heel of palms resting on the keyboard while working;
- working at an angle to the screen, not with the whole body facing it directly;

- glare on the screen, making the worker squint and squirm to read the display;
- screen higher than eye level, plumb vertical or tilted forward;
- frequent awkward reaches for references or other equipment, such as staplers.

FIXING RISK FACTORS

*Hard Fixtures*

Many of these factors can be fixed by adjusting the workstation – or better still, creating an adjustable workstation that can easily be adapted to fit more than one worker during the day. Movable task lighting can reduce glare, and rearranging the workspace (or adding new shelves) can put frequently used equipment or references within easy reach.

Usually the first big change is to the chair. Five spokes with castors, not just four, have become standard equipment because they offer greater stability. Gas-lift chairs are available now that can be raised or lowered with a quick press on a lever, and offer snug lower-back support, cushioned or rounded edges that allow the thigh to hang over, and perhaps armrests for moments during which arms are not on the computer. An adjustable chair with a footstool can to some extent compensate for a too-high desk, putting the worker's face at a comfortable angle to the computer screen and allowing elbows to bend at a comfortable angle to the body.

'There is no absolute right or wrong in chair design,' offered Jim Dixon, OHS project manager with the Canadian Standards Association. 'The right chair depends, to a large extent, on the type of work being performed. As a general rule, however, office chairs should be adjustable without the help of the maintenance department.'[3]

Actually, chairs have become controversial, with a debate raging between the piano-stool advocates and the recliner buffs. Some jobs (cashiering, drafting) and some workers strongly favour sit/stand workstations, with a stool or a bar that folds down for sitting and folds up out of the way to free the worker for stand-up action. New on the market are saddle seats, which leave legs hanging freely to the side – an arrangement that may be more comfortable for men's laterally facing hip joints – and recliners on castors, which can slide under and out of workstations. Balans-style chairs, with the back unsupported and the body's weight leaning against the knees, still have their adherents, too.

Posture and wrist position are partly matters of a worker's own choosing, assuming that desk and chair are suitable. Some say that posture reflects every injury our body has ever received, every bad habit, every bit of social conditioning or self-conciousness or sex-role expectation that we have ever absorbed. Also, when a worker has been sitting in the poke-chin position for years, there's a fair probability that her pectoralis minor muscles have shortened, pulling her shoulders forward, and they won't release without deliberate stretching or massage. So, as Barbara Silverstein pointed out, it's not as simple as having 'the posture police' circulating and reminding workers to sit up straight. With a little MSI-prevention education, most workers will remember to avoid egregiously injurious habits, such as sitting on one leg or lounging with feet up on desk and keyboard in lap, entering data.

However, often people are not aware of how they work, or how they look while they're working. They need lessons in using their bodies efficiently (such as Feldenkrais or Alexander work; see chapter 3) or they need to have an expert watch them (or a videotape of them) at work, in order to spot how their work habits could be hurting them. In an article for *Smithsonian Magazine*, writer Richard Wilkomir described how Dr Emil Pascarelli watched him type on his laptop and analysed his flaws: 'OK, so what you've got is hyperextension of the little finger on your right hand, and that keyboard is putting both your hands into ulnar deviation, especially the right one, and there's some wrist drop too,' the doctor pronounced.[4]

Likewise, wrist position is partly a matter of will-power, partly training, and is influenced greatly by both the workstation and the design and touch of the computer keyboard. The ideal wrist position is 'neutral,' that is, straight as a rod, neither flexed nor extended, bending neither to the outside nor to the inside. Think of a zipper running up your forearm to your palm – any deviation from the neutral position, and the zipper will stick. People with formal typing training tend to sit up straight and keep their arms out straight in front while they work. Most people using computers these days, however, have taught themselves or have learned from computer tutorials, with nobody watching them to comment on their posture or wrist position.

Dr Susan Mackinnon's patient education package explains that simply working with palms down puts a strain on forearm muscles. When our arms are relaxed, we usually place them in a palms-up position – supinated – or sideways. In these positions, the radius and ulna bones in the forearm remain side by side. When our arms are held palms down, or pronated, the ulna and radius are crossed, and either gravity or muscles have to hold them down.

---

*Alert*

Two groups of people who are being urged to learn how to use computers are at special risk for developing MSIs: children, and persons with disabilities.

- Children typically sit at workstations designed for adults. Once engrossed in their work, they seldom look up or take breaks.
- People with mobility challenges may need specially designed workstations. People with hearing or visual disabilities should be rigorous about taking breaks. Hearing-impaired people who rely on sign language should make sure to protect their hands and stop for stretches when hands start to ache. One impairment is enough.

---

As for how specific keyboards affect wrist position – that is another ongoing controversy. Contoured keyboards are readily available now in a wide variety of styles and formats. Some keyboards bend up in the middle, to avoid the palms-down position.

Some put letters into new positions, for greater efficiency of motion. Some have built-in wrist rests. Some computer users say that the shape of the keyboard matters less to them than the touch, the amount of force required to strike the keys. Long-term surveys on the benefits or disadvantages of particular keyboards are inconclusive. CTD*News* and the Office Ergonomics Research Committee (an industry group) have reported no significant differences in MSI prevention from one keyboard to another.

Wrist rests have come under criticism lately. 'If you're going to use a wrist rest, use it to rest,' ergonomist David Rempel told the *Wall Street Journal*. Resting while typing can put pressure on nerves and force tendons to saw back and forth over the carpal bones.[5] A better approach is to install a keyboard tray below desk level, so that elbows open up and hands descend onto the keys. Also, there's a trend towards tilting the whole keyboard away from the user. The 'negative tilt' keyboard almost forces the worker to keep her wrists straight, especially when combined with a keyboard tray below the desk.

Alternatively, some ergonomists and workstation analysts swear by full forearm supports – not just wrist rests – as humans have a natural inclination to lean on their forearms to support their upper bodies when at rest.[6] Forearm rests relieve static loading on the muscles used to keep wrists up, but (theoretically at least) allow hands to move freely. When attached to chair arms, the forearms rests should swivel with the arm motion. Another method is to install a slant board around the keyboard, so that the user sits back (almost reclining) and moves the arm from the shoulder.

Coming under close scrutiny is the computer's desk mouse. VDT Solutions, an office design consultancy, reported that after four hours of mouse use, grip strength dropped by 18 per cent, flexion by 5 per cent, extension by 23 per cent, abduction (ability to spread fingers) by 60 per cent, adduction (bringing fingers together) by 55 per cent, pronation (turning palms down) by 38 per cent, and supination (turning palms up) by 35 per cent.[7]

The point-and-click motion is quick and easy to do, and probably not hazardous if not performed done too often during the day.

However, the drag-and-drop manoeuvre, or drawing with a hand-held mouse – both of which involve squeezing and sustained contraction – become painful fairly quickly. Alternatives to the standard mouse are coming on the market even faster than alternative keyboards and, again, no one model has been proven to be safer. Indeed, when Deborah Quilter and Emil Pascarelli explored computer stores for their book, they reported that 'we couldn't find a single mouse or trackball we felt was safe to use for extended periods of time.'[8]

Touchpads, trackballs, gyromouses that can be used on the lap or in the air – all have their advocates. In fact, some say that the key is to have a variety of devices and use whichever one doesn't hurt at that particular time – or to program the keyboard to perform many of the clicking operations. One concern is that the tracking device should be positioned close to the keyboard, so that the worker can use it without an awkward reach.

Beyond the computer itself, accessories may also present hazards. Ever put a document on the desk and craned your neck to transcribe parts of it into a new document? Well, occasional craning can be counteracted with neck stretches and neck rubs. People who compare documents all day, though, should be equipped with document holders that present the material at eye level.

Then there's that other essential of modern life – the telephone. A sure and certain way to rouse a neck ache is to sit with a phone tucked into one shoulder, held in place by a bent head, and then to try entering or retrieving information at a computer. Again, occasional dual tasks can be overcome. Constantly reaching for the keyboard with a phone tucked under an ear, however, is asking for trouble. Neck strain is a precursor to hand and wrist pain – and plenty painful enough by itself to make anyone want to avoid it.

*Work Organization*

Perhaps the greatest risk factor, though, has to do with the length of the work period. More and more, VDT risk factor checklists ask whether the worker uses the computer more than four (or even two) hours a day.[9] There's considerable empirical evidence that the more hours the worker is actually at the computer, the greater the probability of an MSI.

---

*Losing – Hands Down*

Dr Susan Mackinnon wrote: 'Individuals who do a considerable amount of keyboarding will keep their forearms in a palm-down position for the majority of the day. This will allow the pronator teres muscle to shorten. When it becomes short and tight, it can compress the median nerve in the forearm. This same palm-down position will also cause two tendons in the forearm to compress the radial sensory nerve, producing numbness over the back of the hand ...'[10]

One common result of a tight pronator teres is a syndrome that closely mimics carpal tunnel syndrome but which will not respond to most CTS treatments – certainly not to a surgical carpal tunnel release, which won't even touch the area where the median nerve is actually compressed.

---

TEN-SHUN!

Whatever entrapment an expert blames, there does seem to be general agreement that holding one work position all day can cause MSI. Most, but not all, office workers have some discretion in choosing which tasks they do and when. And more and more, tasks switch between one computer program and another. Clerical workers, for example, used to have to leave their desks to get new materials, and to deliver and file documents. Now all those tasks can be done on-screen, without getting up.

In addition, some companies have implemented (and many more have flirted with) computer monitoring of employee productivity even though, as OHS Canada's consulting editor Cindy Moser wrote, 'computerized tracking could wreak havoc on an office health, safety and ergonomics program.'[11] Think of rows and rows of data-entry clerks – invoice typists, contract and litigation fillers-in, dictation transcribers – kept at their desks by the invisible chains of keystroke counters. Ten thousand keystrokes an hour, all day, or find a new job.

Moser described two cases of tracking: the first involved Anne, an airline reservation clerk whose supervisor had told her that she spent an average forty-two seconds more per call than the three-and-a-half minutes that the company allowed, although random checks showed her to be courteous and thorough enough to meet the company's expectations. 'Anne is damned if she does, damned if she doesn't,' commented Moser. 'It's the classic stress-producing situation.' And the stress is fruitless, she argued, because computer monitoring has no effect on productivity, as demonstrated by her second example.

Bell Canada tried a monitoring program that kept track of the number of calls each operator handled and compared that number to the office average. Not only were operators stressed out, but '50 per cent of the operators were destined to fall short no matter how hard they tried. That's how averages work.' When the monitoring was discontinued, Bell expected productivity to drop – but it didn't. Moser urged OHS officers to keep files of such examples to use in deterring their own companies from ever implementing computer monitoring.

BREAK TIME

Some companies use keystroke counters in a different way – to remind employees to take breaks every hour, or every few thousand keystrokes. Workers who choose their own programs are more likely to go along with the routine. 'Some people just hate them,' said ergonomist Ilene Stones. 'They disable them as soon as they can.'

Many newspapers have installed take-a-break programs that pop up and remind journalists and editors to go shake out their hands. These don't always have the desired effect. Barbara Silverstein studied a newsroom with this kind of program and found that 'people would work even harder, really banging away at their keyboards, when they knew the program was about to pop up.'

Behind the software that blacks out screens automatically for a few minutes every hour, behind the little seventh-inning stretch cheerleaders that bounce onto some screens, lies a new conventional wisdom. The *Wall Street Journal* cites the new expert advice: 'Typists should rest their hands about 10 minutes an hour.' In this article, Stephanie Brown (CEO of Ergonome software publishing) reported that experts suggest 'using short frequent breaks rather than long ones ... Many recommend microbreaks; dropping hands to the lap for just five seconds can be helpful.'[12]

Along with microbreaks, most experts recommend regular stretch breaks during the day. Some workers are embarrassed to stretch extensively at their desks, but stretch breaks can be as simple as using one hand to flex and extend the other – as far forward as it will go, then as far back as it will go easily – and then changing hands. Shoulder rolls, neck stretches, and wrapping arms around shoulders (as though giving yourself a hug) are great ways to loosen up the shoulders and neck, which are often implicated in upper-extremity MSIs.

After MSI outbreaks, some companies (such as the *Los Angeles Times*) have set aside special rooms for stretch breaks and other MSI-prevention activities.[13] Some companies already provide fitness facilities in-house or access to nearby health clubs – after an outbreak, they might strongly encourage staff to work out more often. Some manufacturing plants have experimented with mandatory exercise sessions at the beginning of the workday, but the results have generally been disappointing.[14] Exercise under duress can do only so much to relieve stress. A U.S. National Institute of Occupational Health and Safety (NIOSH) study of exercises recommended for VDT operators found that one-third were conspicuous, half of them would disrupt the work routine, and several actually had the potential to cause physical problems.[15]

---

*Warm-Ups*

Stephanie Brown's *The Hand Book* contains instructions for warming up, exercising, and massaging hands before, during, and after computer use. Drawn from her background as

# Staring at the Screen: Computer Workstations 115

> a classical pianist, Brown's methodology combines movement retraining and self-care with evocative descriptions. 'The main thing to remember is that typing actually takes very little effort,' she wrote. 'I'm sure you've seen people who type by slamming their fingers into the keys, as if they're nailing them to the keyboard or destroying little bugs lurking underneath ... when we think of control, speed and accuracy, we think we need to tighten up – as if that extra effort will produce those results. The opposite is true.'[16]

*Spicing Up Work*

In the *World Health Forum* journal, Professor Choon-Nam Ong fingered stress as an MSI issue with VDT workers. 'Clerical personnel who use visual display terminals generally have little work involvement, job autonomy, self-esteem or supervisory support,' wrote Ong, 'and tend to have more health problems than professional staff, apparently because they have higher stress levels at work ... the problems are clearly related to feelings of lack of control, lack of variation, rigid procedures, higher production standards than in traditional office work, and constant pressure for performance.'[17]

> *All Right, Break It Up*
>
> When monotony causes health problems, the solution is obvious: variety. 'Vary the work tasks of office employees,' Ilene Stones and Wendy King advised in *OH&S Canada*. 'Break up VDT work with non-VDT work that places different demands on the body.' Also, 'provide good work/rest schedules and encourage workers to take their scheduled breaks. Encourage workers to stand up, move around and change mental activity during rest breaks.'[18]
>
> The Public Service Alliance of Canada manual concurs: 'Jobs must be designed to reduce both monotony and prolonged physical exertion,' it advises. 'Variation and self-determination are key elements in reducing particularly

> repetitive strain injuries, stress and visual disorders associated with VDT use.'[19]

As one contributor to the SOREHAND listserv commented, modern offices are computer-driven. It seems that the more money employers spend on computers and technological equipment, the more the work gets rearranged to make tasks easier for the machinery – even if that means the work is harder for the human resources involved. Computers make good servants but poor masters. Yet, more and more workers find themselves serving machines. And when health problems develop, often the first solution offered is more technology.

PICARD TO SICK BAY

Dr Brendan Adams suggested that computer-related MSIs would soon be history, because his work with a telecommunications company convinced him that 'we're going to true voice recognition at the speed of light.' Shades of *Star Trek*! Imagine being able to sit down at a desk (or pace freely around an office) and call out, 'Computer, send e-mail invoices to all our monthly accounts. Take the names from the Sojourner file and the amounts from the Charlotte file.' And poof, it would be done.

Unfortunately, true voice recognition seems to be developing more slowly than the technowizards promised, or than MSI sufferers have hoped. And the slow, stacatto style of speech that is required by older software has proven risky, too. Some users have developed painful wasting conditions, such as laryngeal nodes or strained vocal cords. 'Those affected develop chronic hoarse throats and may lose their voices altogether unless they can learn to recondition the muscle through breathing and voice exercises,' according to Mary Gooderham in the *Globe and Mail*.[20] 'Speech therapists and throat specialists call it vocal abuse, and they are becoming increasingly alarmed.' Such problems may be avoidable – new baseline voice tests can identify people who are poor risks for voice recognition – or soluble with the latest voice-recognition software, which is still hovering at the $3,000 price range, and requires Pentium 200 mh or larger.[21]

Nor does virtual reality (VR) seem to offer pain-free alternatives. At least, not yet. Readers who have watched the movie *Johnny Mnemonic* or the television series *Tekwars* may be struck by the metaphor of 'flying' into VR, using their whole body, free from keyboard or mouse or any connection besides electrodes on their fingertips. So far, that vision remains in the realm of fiction. According to a paper on safety considerations in VR, 'Repetitive Stress Injury is a somewhat common result of extrended VR use ... The primary cause of this condition is the rapid carpal and meta-carpal movements used in joystick and keyboard controls.'[22]

BEYOND EQUIPMENT

By themselves, one-size-fits-all technological fixes cannot prevent typing injuries. The key is to supply equipment that fits the worker – which means that the worker has to know her or his own body well enough to recognize what helps and what hurts.

'Good equipment by itself is not enough,' wrote Paul Linden of the Columbus Center for Movement Studies. 'A bad chair will make good, comfortable sitting posture difficult if not impossible. But even a good chair won't help much if you are abusing your body through muscular tension and awkward posture. Sitting right on the wrong chair won't work, and sitting wrong on the right chair won't make you comfortable either. Effective movement technique and proper equipment are both necessary in order to compute in comfort.' This belief (grounded in a PhD in physical education) led Linden to write 'a whole book about how to sit,' as he has joked in e-mail. Actually, *Compute in Comfort*[23] includes a whole variety of lessons in movement retaining, as well as stretches and exercises for work breaks, and a discussion of safe computer equipment.

Computers might just be the most useful tool since fire – but we design our workplaces and homes to keep fire in its place. The more that work is redesigned to fit what computers can do, the less the work fits what people can do. The way to reverse our present momentum towards workplaces that are friendlier to computers than to people is through applied ergonomics.

# 8
# Fitting the Jobs to the Workers

VOLUNTEERS, PLEASE TAKE TWO STEPS FORWARD

When a repetitive strain injury epidemic arises suddenly and dramatically, an employer may be forced to take action. For instance, soon after Cuddy Foods opened a new poultry processing plant in London, Ontario, in 1987, president Robert Cuddy realized something was wrong. According to *Canadian Business Magazine*,

As many as 44 workers *per month* became disabled, some permanently. Some of the injured workers were as young as 22. Slips and falls accounted for a few of the injuries. But 73 percent were victims of ... repetitive strain injury ...
Cuddy's Workers' Compensation Board penalties reached $850,000 for the three-year period ended 1988 ... Local doctors were so fed up with the number of Cuddy Food casualties trailing into their offices that many refused to even speak to the company's physician ... Says Cuddy, 'We were dealing with the walking wounded. And we created that ourselves.'

Cuddy Foods finally called in an ergonomist to redesign the workplace, at a cost of more than one million dollars. Not only did injuries fall to zero, but productivity rose, too. Still, retrofitting a brand new state-of-the-art assembly line was a tough business decision, and one that could have been avoided if the potential for MSIs had been taken into account in the planning stages.[1]

Nor is Cuddy Foods unique. CTD*News* frequently spotlights case studies of ergonomic adjustment. A 1995 special report on best ergonomic practices, for example, said that Red Wing Shoes of Red Wing, Minnesota, cut its lost-time days by seventy-nine per cent in five years, and consequently reduced its WCB premiums by 70 per cent. AT&T's San Diego plant decreased days lost due to MSIs from 298 in 1990 to none in 1993, and consequently WC costs plummeted from $400,000 in 1990 to $8,600 in 1994. General Seating, of Woodstock, Ontario, cut its reported injury rate by almost three-quarters in just one year, from 1,136 in 1993 to 335 in 1994.[2]

NOW, WATCH CAREFULLY

An ergonomics program has to be monitored to ensure its effectiveness. Occupational health nurse Kathleen Buckheit talked about her work at a furniture plant in North Carolina. 'One week there were five sore wrists on one line,' she said. The program's emphasis on early intervention encouraged workers to come to the nurse's office when they first started to ache. Buckheit applied first aid (ice), and 'then you go look at the job, to see what's hurting people.'

The job involved used a 'breakaway' screwdriver – first with a pistol grip for the side screws, then straightened out for the top screws, so that the worker's wrist was always straight. 'Some of the workers discovered they could save a second and a half if they didn't use the breakaway. They were using the pistol grip on the top screws, and getting a few seconds to chat.' Buckheit talked with the workers about keeping their wrists straight. She also talked with management – and the work was redesigned so that one worker put in side screws and the next screwed down the top.[3]

'The ... case studies make it clear,' according to CTD*News*, that controlling workplace cummulative trauma disorder 'involves more than just a one-shot reactive effort ... employers must be progressive if they are to prevent CTD outbreaks – even if their major ergonomics effort is education.'[4]

> *Defining Ergonomics*
>
> The Board of Certification for Professional Ergonomists (BCPE) defines ergonomics as 'a body of knowledge about human abilities, human limitations and human characteristics that are relevant to design. Ergonomic design is the application of this body of knowledge to the design of tools, machines, systems, tasks, jobs and environments for safe, comfortable and effective human use.'[5]

Two other terms are sometimes used interchangeably with ergonomics. These are 'human factors' and 'biomechanics.' In fact, according to the ErgoWeb website, these are two distinct branches of the profession: 'One discipline, sometimes referred to as "industrial ergonomics," or "occcupational biomechanics," concentrates on the physical aspects of work and human capabilities such as force, posture and repetition. A second discipline, sometimes referred to as "human factors," is oriented to the psychological aspects of work such as mental loading and decision making.[6]

That's an ergonomist's definition. Cathy Walker, Director of Health and Safety for the Canadian Auto Workers' union, suggested that workers use a much broader definition: 'To some degree, ergonomics encompasses absolutely everything within health and safety,' she said. 'Our view of ergonomics goes far beyond the very narrow industrial engineering definition that corporations put forward. It's the whole definition of work – things like how work is done, pace of work, even overtime and vacation.'[7]

GETTING A HANDLE

The British Columbia WCB Draft Ergonomics Regulations highlighted the central premise of the discipline in big, bold typeface: *'Ergonomics fits the job to the worker.'* But the devil is in the details. A workstation that fits one worker perfectly might need some changes to fit the next. As well as the physical set-up, ergonomics examines job duties, starting with FFDP – force, frequency, posture, and duration of repetitive movements. Vibration, twisting, and workplace temperature are also key factors.

Accurate illness-and-injury record keeping should help identify problem workstations. As well, various codes of practice provide checklists of signal risk factors, such as:

- performing the same motion or motion pattern every few seconds for more than two hours at a time;
- a fixed or awkward work posture (e.g., overhead work; work that involves a bent wrist or stooping posture) for more than a total of two hours;
- using vibrating or impact tools for more than a total of two hours;
- forceful hand exertions for more than a total of two hours.[8]

Along with these assessments, 'It is important to include workers in the process of identifying, assessing and controlling risks,' according to the BC WCB. 'Workers often know best the activities and tools that contribute to their pain, and have practical suggestions about how to eliminate or minimize the risk of adverse health effects.'[9]

---

*Bonus Time*

Of all the occupational diseases seen at the Toronto Occupational Health Clinic for Ontario Workers 'Ergonomic conditions are probably the easiest to correct because of workplace design,' reflected executive director John Van Beek. 'If you correct an ergonomic problem, what you've done is not only reduce any MSI problems that might be occurring, but you've also increased productivity levels. As soon as people pick up on that, they're sold.'[10]

---

CAN WE TALK?

Consultation with workers is a crucial step – but it's too often ducked or overlooked. For employers to ask workers whether their jobs are hurting them requires a real leap of faith. Such a question might be construed as an admission of culpability, which could be costly if any employees later took legal action. Likewise, for employees to tell their employers which tasks cause them pain requires them to trust that their jobs are secure – even if they complain – and that the bosses are not just looking for lame ducks or troublemakers to let go. Further, workers need more than just an opportunity to make contributions. Frequently, they need to be startled into awareness.

'At [the electronics assembly plant], I purposely started a panic,' said kinesiologist Greg Hart. 'Especially with the young women, I'd ask them, do you plan to have children? Do you want to be able to pick them up? That scared them into reporting all their aches and pains.'

'One of the things that you'll see with Ford, and one of the things that we value a lot, is our relationship with the union,' said Hank Lick, Manager of Industrial Hygiene for Ford Motor Company, at an interview during the Managing Ergonomics conference.

I think you can say that we jealously guard that. What we have is many years of building a relationship of trust. So that if the management people at the health and safety end of the business say that it's a problem, or it's not a problem, the [United Auto Workers'] end of the business tends to agree ... The UAW put very good people in charge when we started talking about ergonomics ...

We spent a lot of time teaching what we call ergonomics committees, which consist of eight people, four from management and four from labour. We taught them basic ergonomics, how to prioritize, how to communicate with each other, and finally how to document. What we saw was that it's more than just worker participation; they have to know how to participate.'[11]

PLEASE LISTEN CAREFULLY

The worst way to eliminate hazards is just to call in an engineering company and have the engineers go by the numbers, rebuilding workplaces to pre-set heights, weights, and widths. Standard sizes just do not work with the workforce of the 1990s. Another common error is to commission the purchasing department to replace chairs (the most common ergonomic adjustment), desks, worktables, and other workstation equipment without inviting the affected workers to help select the new equipment.

You may have seen business stories such as this one from the *Calgary Herald*: 'Many people are in full denial when it comes to ergonomics in the workplace, says [a consultant],' who talked about meeting resistance from workers when she tried to reorganize their workspaces or replace their equipment. She talked about 'university graduates who refuse to wear a telephone headset to avoid neck and shoulder pain – preferring to cradle the telephone handset between their shoulder and their head while typing.'[12] In contrast, the consultant said, companies are more ready to listen.

Good intentions are not enough. Workers don't like to have changes imposed on them, even for their own good. What people don't like, they will not use. Moreover, if their employer does not bother to discuss the new equipment with them in advance of the purchase – to explain what the new workstations are intended to do, and ask for their ideas about achieving lower injury rates – workers may hesitate to voice their problems with the new setups. As health and safety officer Frank Hamade has noted, 'a worker can have the best and most expensive equipment in the world, but if it is the wrong equipment or it is not set up properly, an injury can occur.'[13]

ERGONOMICS AND WOMEN

Women are at greater risk of having to work with the wrong equipment simply because most workplaces are built using anthropometric measurements. That is, tables, chairs, desks, conveyer belts, hand tools, and workstations are all designed to accommodate the so-called average man – as visualized from a set of measurements taken from U.S. army recruits at the beginning of the Second World War.

The result is that 'a traditional work surface in the United States is 29 to 31 inches [74 to 79 cm] high, ideal for a man who is 5 feet 10 inches [178 cm] tall but totally inappropriate for a woman of 5 feet 1 inch [155 cm], who requires a work surface approximately 23 to 25 inches [58 to 64 cm] in height,' according to Dr Linda H. Morse and nurse practitioner Lynn J. Hinds, in an article published in *Occupational Medicine: State of the Art Reviews*. They called for research to develop new anthropometric and biomechanic data that can accommodate a diverse workforce, comprising not only women but also members of ethnic minorities (such as Hispanic and Asian people) who tend to be smaller than 5 feet 10 inches [178 cm]. 'Failure to do so,' they warned, 'will perpetuate the rising rate of injuries due to ergonomic hazards.'

Hinds and Morse suggest a number of low-cost methods for adapting tools and workstations so that people of various sizes can use them comfortably and safely. But their main message is that the field of ergonomics needs to pay more attention to *who* the new workers are. 'Studies of musculoskeletal problems in women workers have not kept pace with the dramatic increase in women in the workforce,' they noted. 'In fact, this lack of studies has been cited as evidence that no problems exist.'[14]

Dr Barbara Silverstein, Lawrence Fine, and Professor Thomas Armstrong suggested in a 1986 paper that even when gender is identified as a factor, the identification is erroneous. In a survey of six companies (subsequently cited in practically every medical or ergonomics journal that publishes articles on RSIs), they established that workers' likelihood of developing RSIs can be predicted by the force and repetition used in their jobs. The study did not look specifically at posture, but the authors noted that 'wrist postures required on a job are often determined by the height of the work station with respect to the location of the worker. A tall man may use less wrist flexion or ulnar deviation than a woman (or a shorter man) performing the same job. In this example, what may be assumed to be a sex difference would in reality be a difference in working posture.'[15]

## Fitting the Jobs to the Workers    125

TEAMWORK, TEAMWORK

Since MSIs are multifactorial in origin, most successful attempts to alleviate them are also multifactorial in approach. That is, smart workers and employers realize that they have to work together as a team in order to identify and eliminate hazards for MSIs in their shared workplaces.

The Washington State guidelines suggest that for small businesses, the MSI team should include workers or union representatives, managers and supervisors, maintenance staff, safety and health personnel, and purchasing personnel. (In some small businesses, staff have overlapping duties and one person may represent more than one category on the committee.) Larger businesses need larger teams – all the personnel above, plus engineers, human resources personnel, healthcare providers, and an ergonomist. Multidepartment businesses may need several ergonomics teams, perhaps one team for each department or section.

'Ergonomics programs typically have these four elements,' according to the guidelines: worksite analysis, hazard prevention and control, medical management, and training and education. Because such programs are usually set up in response to acute situations, they tend to start by focusing on existing hazards and encouraging workers to report incipient injuries in the early stages, before they crash. 'As the program develops, and the acute situation eases, the focus can shift to prevention.'[16]

STOP THE BLEEDING

Often, the first step is to let workers know that they have options in medical treatment. At the power supply manufacturing company investigated by the California OSHA (discussed in chapter 5), the first thing the ergonomics team did was to intervene in medical management of injured workers, who were being directed to surgery and then right back to unmodified workstations. Only after that could the team concentrate on fixing the built environment and the tools where workers did their jobs.

Jim McCauley, an ergonomist with Perdue Poultry, emphasized that medical management was an essential component of Perdue's program, to the extent that the company consults with physiotherapists about which movements on the line might be part of a worker's PT regime. That might sound harsh, but every week a worker is away reduces the likelihood of her return. Perdue also requires a second opinion any time a doctor recommends carpal tunnel release. 'Doctors around here used to do both hands at the same time,' said McCauley. 'We don't allow that any more.'[17]

This last comment is in line with recommendations developed by occupational health nurse Pat Bertsche, formerly a member of the OSHA and ANSI ergonomics committees, and since 1996, manager of the new Ohio State University Institute for Ergonomics. According to an article she co-authored for *AAOHN Journal*, 'while carpal tunnel release surgery has been reported to be 80 percent to 90 percent effective in decreasing or relieving the pain, its effectiveness in returning employees to their original jobs is 40 per cent to 50 per cent at best ... All employees scheduled for carpal tunnel release surgery should have: 1) their treatment program reviewed to assure that conservative therapy has failed, and 2) a second opinion to corroborate the need for surgery.'[18]

*Early, Earlier, Earliest*

But early intervention also means coming right out and asking workers about pain in their arms. Oddly, workers do not always welcome such an initiative, at least, not at first. Workers get used to the workplace routine. They develop inertia. Walk through most workplaces and ask workers to name their most urgent work concerns, and very rarely will they volunteer thoughts about health or safety. Usually the majority will name wages (or salary), job security, work hours, and maybe benefits as their top issues. In the United States, medical insurance is liable to be somewhat higher on the priority list.

The fact is that, whether or not an OHS agency is knocking at the door, an employer needs foresight and patience in order to initiate an MSI prevention program. Doing it without pressure from an enforcement agency may allow employer and workers more time to get it right and promote a less adversarial atmosphere. Since close cooperation is essential for identifying ergonomic hazards, and trust is a component in accepting workplace changes, managers can make their own work easier by introducing ergonomic changes in a more relaxed manner.

Now, walk through that workplace again, and *ask* employees about their aches and pains. After reviewing health and safety records, that's one of the first activities that OSH committees (or MSI consultants) undertake in a new workplace. Workers who might never connect their sore neck or aching hands with their job – not enough to name health as a job issue – often respond candidly to direct questions about whether they're in pain and where and how often. The Institute for Work and Health has developed a thirty-item questionnaire to identify disorders in any part of the upper limb.

The magnitude of potential problems thus revealed can be breathtaking. At the 1997 Managing Ergonomics conference, Barbara Silverstein cited a 1988 survey of U.S. workplaces that found '20.7 million people had hand discomfort that had lasted more than one day.'[19]

In British Columbia, a questionnaire that the Women and Work Research and Education Society distributed in 1993 through unions for auto workers, government employees, and food workers found that 53 per cent of respondents reported hand or wrist pain, and 65 per cent reported arm or shoulder pain. Thirty-nine per cent experienced the early symptoms of numbness and tingling.[20] In 1996, a study conducted by the Canadian Auto Workers' Union and McMaster University found that fifty-five per cent of workers at the Big Three auto plants are working in pain for much of the time, and almost two-thirds are constantly tired or tense.[21]

Employers don't like this approach. They claim (with some justification) that everybody has aches; they can point to a survey by Louis Harris and Associates that found two-thirds of the full-time workforce suffers from pain on the job, but only half the pains are job related.[22] Asking workers to say where they hurt, they claim, is an open invitation to misunderstanding, misrepresentation, and malingering.

However, when a workplace or an occupation has been identified as a high-risk area, such direct questions help pinpoint the workers who may be in need of help – and confirm or dispel researchers' hypotheses about the safety of a workplace. When more than half the workers in a given work area report musculoskeletal pain that has persisted more than a week and has occurred more than once in the last six months, the employer is probably looking at a serious problem.

In many companies, management snoozes through the early warning signs of developing MSIs, perhaps misses the middle-stage signals because of downsizing or corporate restructuring, and wakes up with a start when the human resources or employee relations department figures show high rates of sick leave and employee turnover.

*Don't Ask, Don't Tell*

The buzz from Calgary office towers is that major Canadian companies seem to be dealing with MSIs on an ad hoc basis. Bosses recognize that some of their workers are injured and authorize time off, ergonomic equipment, and other accommodation without ever notifying their own health and safety departments, or filing an injury report. An injured office worker, especially if she is a professional, can pretty much write her own recovery plan, without ever acknowledging any impairment.

While such arrangements might seem generous, they do have a few flaws, including inefficiency, isolation, and policy vacuum. Each worker has to research the problem from scratch. Each boss has to determine how much flexibility is reasonable, without affecting other workers. Most of all, lack of statistical data means that neither the company nor the industry knows what the incidence really is, much less what ergonomic adjustments are effective. An ad hoc approach means that work injury becomes a personal problem, despite the wealth of evidence that a team approach is the most effective way to implement ergonomics.

Fitting the Jobs to the Workers    129

WORKSITE ANALYSIS

While the medical management team deals with workers who have reported injuries and surveys others about possible injuries they haven't yet reported, other members of the team may take checklists and perform a facility walkthrough, looking for ergonomic hazards. They closely observe work procedures in order to identify postures, tasks, tools, and workstations that might cause problems.

Expertise helps, but plenty of guides are available to help even beginners spot potential problem areas. ErgoWeb offers both a general risk factor summary checklist (applicable to a variety of worksites) and specific checklists for warehouse work, small-assembly work, and computer workstations.[23] The Occupational Health and Safety Administration in the United States has published codes of practice for meat-packing plants and for materials handling, also available from ErgoWeb. The National Institute for Occupational Safety and Health offers guidelines and free consultations. The risk factors described in chapter 1 could serve as a preliminary checklist, in a pinch.

Workbooks from the Ontario Workplace Health and Safety Agency feature musculoskeletal injuries charts that alert worksite analysts to links between common MSI injuries and risk factors specific to various jobs. For instance, under bursitis, 'awkward reaches' may occur in the construction industry among bricklayers and rodmen, in manufacturing when workers have to pull products along the belt of an assembly line or packing area, in processing and assembling among textile workers who have to reach overhead to remove spools, and among clerical workers when they reach up or to the side in order to use a computer mouse.[24]

The BC Draft Ergonomics Regulations contain several checklists for analysing job demands to bring them into compliance with the code of practice. The code looks at:

- physical demands of work, including force required to lift, lower, push, pull, or grasp;

- repetition and duration;
- static or awkward postures;
- workstation design, including reaches, working heights, seating, and floor surfaces;
- characteristics of objects handled, including object size and shape, tool and equipment handles, local contact stresses, and vibration from hand tools;
- environmental conditions, such as temperature, lighting, and whole-body vibration;
- work clothing and protective equipment; and
- work organization and pacing.[25]

DETAILS, DETAILS

You could take a slew of ergonomic guidelines, stack them up and shuffle them, deal them out randomly, and they would probably all end up listing approximately the same ergonomic hazards that they started with: FFD (force, frequency, and duration of repetitive motions), posture, vibration, and working conditions, especially temperature. Where they differ from program to program, is in stipulating how to control hazards once they've been identified, in order to prevent future injuries. Yet, all these programs achieve much the same goals.

For example, the first draft of the BC ergonomics regulations stipulated precise physical measurements for workstations, such as requiring that worktables for standing work be 85 to 110 cm tall, or 5 to 10 cm below the worker's elbow height; the depth of a seat for an office chair was to be 38 to 43 cm, and the compressed seat height 38 to 52 cm. By contrast, the regulations that British Columbia finally enacted were more general, similar to the Washington State guidelines which say only, 'Work surfaces should be at the proper height and angle for the individual worker's size and tools and equipment used. They should permit neutral postures and be adjustable, especially where different kinds of tasks are performed or the workstation is shared.'[26]

Yet another agency, which sometimes has been forbidden to publish or promulgate ergonomics regulations, would require only that 'the employer shall control each problem job, and shall ensure such control is maintained, by implementing engineering controls or administrative controls which the employer demonstrates are effective in controlling the job.' Attached to this agency's guidelines would be a list of non-mandatory guidelines that employers and workers could use as benchmarks for evaluating new or rearranged workstations.[27]

The problem is that an ergonomics standard is damned if it does get specific – because then employers can complain that it's nit-picking – and damned if it doesn't, because then employers can complain that they don't know what they're expected to do. British Columbia's guidelines in fact were held up for public ridicule on the television program *W5* precisely because they were so detailed. Host Eric Malling rattled off exceptions to every proposed rule and finished by deriding the box in which the WCB's supporting information was delivered, which he said exceeded the proposed ergonomics standard's guidelines for lifting. (It's hard to tell how he worked this out from the tables, which factor in how many times a day the worker would have to lift the weight.)

Employers don't like to have new methods imposed on them any more than workers do. With that common starting point, the answer seems clear. Teamwork is not only the essential *first* step in identifying ergonomic hazards; it is an essential component at every step towards preventing future MSIs. When every person who is affected by proposed changes has had an opportunity to comment on them in advance, that improves the acceptance rate.

---

*It's Everybody's Baby Now*

'The kind of thing that needs to happen is not just to have an engineers come in and say, oh, we can do this that and the other thing and everything will be fine,' said Dr Barbara Silverstein. 'What you really need to do, in my view, is first to provide workers with enough information themselves so that they know what it is you're trying to do and why. Then you talk with them about reducing risk factors or multiple risk fac-

> tors, and you listen to their ideas about how they would go about doing it – bring in the engineers, or the maintenance people, talk together about how to do it, test it to see if it in fact does what you think it's going to do, and as it begins to work, put it into production. Then you have workers who have bought into [the change], you've got engineers who have bought into it, you've got supervisors who have bought into it, and it's got a much higher likelihood of success.'[28]

## FROM ANALYSIS TO ACTION

Identifying hazards may actually be the easy part. The fun begins when whoever is in charge starts to suggest solutions. Management wants to keep expenses down, workers want to maintain their earnings and favourite workstations, unions want to protect seniority and wages, everybody's concerned about hours (though from different perspectives) – getting to agreement can be an emotionally draining process. Sometimes an outside consultant can ease negotiations by acting as an impartial third-party – and also by virtue of his or her experience with a range of compromise solutions.

For instance, if management complains there isn't time between shifts to adjust workstation heights, the solution may be to put workers on scissor-lifts. Management may want workers to begin every day with an (unpaid) exercise period; if workers balk, the solution may be to incorporate five-minute stretch breaks two or three times a day. Job rotation cannot just be imposed in workplaces where unions and management have already agreed on strict hierarchies of duties and wages – persuasive negotiation is required.

Kinesiologist Greg Hart said that part of his job in evaluating worksites is to balance two opposing requirements: 'Ergonomics is technologically centred,' he observed, 'but work needs to be person-centred.'

Of course, the size and scale of the enterprise also enter into design decisions. During risk manager Gail Sater's video presentation on changes at Red Wing shoes, she talked about job rotation and how shorter workers had their own platforms (of hard foam, with plywood sides) which they carried with them from workstation to workstation.[29] By contrast, United Auto Workers' project ergonomist Lida Orta-Anes said in an interview that the UAW tries to avoid workstation platforms 'because the worker might fall off. We'd rather adjust the height of the line [conveyor belt].' Since auto plants re-tool fairly regularly, and since workers tend to stay with the same plant, lowering the line is not as difficult for employers as it might seem.[30]

Despite the way that workplaces are more and more technologically centred, many ergonomic fixes are fairly low-tech. Barbara Silverstein suggested that a meat cutter might avoid the common problem of 'trigger finger' by using a sling to keep hold of the cutting knife while still allowing middle, ring, and little fingers to stretch out frequently during the day. Or an electronics assembly worker might have forearm rests, thus avoiding static loading in forearm muscles, which could lead to nerve entrapment. Sater's video showed workers resting their arms in slings hung from the factory ceiling, which allowed them to handle heavy workboots with minimal strain. 'One of our workers sewed these,' she said, 'until we could find a supplier.'

The United Food and Commercial Workers' 'Ergonomics Program' fact sheet offers case studies, some with photos, to illustrate ergonomic conversion. Dull scissors and knives are mentioned several times as hazardous to workers' hands, and the UFCW recommends other ways to cut. One meat-packing plant replaced vibrating knives with a machine that uses compression to remove meat from bones. 'In another meat-packing plant, workers had to reach up or bend down to trim contamination from the beef carcass. Now, hydraulic adjustable stands move workers up and down. This allows workers to better reach the meat without reaching or lifting.'[31]

*Worksafe*, the newsletter of the Manitoba Workplace Safety and Health Branch (WHSB), reported a case study involving a local manufacturer of packaging materials that produces nine million kilograms of finished product each year – a business that has grown steadily since it was founded in 1977. In the early 1990s the company saw a sharply increased rate of MSIs. In 1994 the company asked for help, and the WSHB supplied ergonomist Susan Wands.

The company already recognized that 'many of the existing practices ... were based on the worker adapting to the workspace, with no provision for size or physical ability.' At one station where manual lifting was required, the company installed a new roller belt so that the product could be rolled onto pallets, and modified an overhead hoist. At another station, the company installed an adjustable indexing table and a scissor-lift to bring the product to a comfortable working position. When these first adjustments proved helpful – and acceptable to workers – the company expanded its ergonomics effort to redesigning other machines in both areas.

In the first department, where the company had recorded eighty-one strain injuries and lost more than 3,500 hours from 1991–4, there were only nine minor strain injuries and sixty-eight lost hours in 1995. In the second area, which had seen thirty-four strain injuries and 6,700 lost hours during the same four-year period, ergonomic changes helped reduce the injuries to five and lost hours to sixty-two, in 1995. Overall, 'WCB claims fell to their lowest-ever levels in 1995.'[32]

*Simple Fixes*

CTD*News* offers case studies regularly. 'The ideas aren't complex or expensive, and are usually developed first by plant personnel,' says the introduction to CTD*News*'s *Ergonomic Resources Guide*. Take the case of an operator of a bench-mounted machine who spent all day bent over at the neck and wrists. He mounted the back legs of the bench on a small block of wood, which angled the work up to meet the operator, whose neck and wrists could work in a straighter, more neutral position. Cost: nothing.[33]

*Tailor-Made*

Although many experts believe that preliminary checklists of

## Fitting the Jobs to the Workers 135

workplace hazards can be generic (some employers dispute this), worksite evaluations and ergonomic adjustments have to be site-specific, and worker-specific, too. Finally, changes must be reviewed periodically because workers change, job demands change, and sometimes the workplace is retooled or upgraded. That's why the team approach to ergonomics has to be ongoing.

Some jurisdictions require occupational health and safety committees in every workplace. According to the British Columbia WCB, joint health and safety committees are mandatory for all workplaces with more than twenty employees in British Columbia, Manitoba, Nova Scotia, Ontario, Quebec, Saskatchewan, and the Yukon, with duties that may include inspecting the workplace, investigating accidents, and ensuring that workers have health and safety training. Unions often negotiate for OHS committees in their collective agreements. Even without legal or contractual obligation, employers could do worse than to initiate OHS committees themselves.[34]

Auto workers and auto manufacturers have been at the top of the risk list for MSIs. Perhaps as a consequence, the auto industry is in the forefront of the ergonomics evolution. As Andrew H. Card, Jr, president and CEO of the American Automobile Manufacturers' Association, said in his keynote address to the 'Managing Ergonomics in the '90s' conference in February 1995:

... Ergonomics is really about people – creating an environment that fosters safety and productivity ... America's car companies have learned a lot from ergonomics. First and foremost, ergonomics programs must be customized for individual workplaces. Our companies have not tried to implement cookie-cutter programs to be used at all facilities, but are taking the time and effort to evaluate the unique needs of each facility.

Second, a well-designed ergonomics program can reduce some of the costs of doing business. It's much easier – and less expensive – to prevent problems than to fix them after they've occurred.

Third, America's car companies have found that ergonomics can improve not just health and safety, but also product quality and productivity. A well-designed ergonomics program will help workers do their jobs more efficiently, which also has a positive effect on both the product and the company's bottom line.'[35]

Ford's program in the United States, for instance, truly is 'a world-class program,' as Hank Lick put it. Epidemiologists Gordon Reeve and Susan Pastula described Ford's data-collection system: 'the Ford Motor Company maintains a Health Data Analysis system to access occupational injury/illness data that include 55 of its U.S. production facilities ... [which is] linked electronically to several other data systems within the Company ... [including] the payroll system, to allow analysis by department and job ... [and] with Workers' Compensation data ... A cost per case can then be generated from the HDA system,' instantly, right up to the date of the request.

Alas, the system stops at the U.S. border, according to a report issued jointly by the Canadian Auto Workers and McMaster University on 3 June 1996. 'In an era of lean production designed to speed up work and regiment every second of every minute with stop-watch monitoring,' said the CAW, 'workers and their families are paying a grave price.'

The results of more than 2,400 surveys (with a 57 per cent response rate) showed that more than half (55 per cent) reported they are working in pain for much of the time, and that three-quarters (75 per cent) said they are working at an excessive pace or are overloaded. Almost two-thirds are constantly tired or tense, and 78 per cent don't believe they can keep up the pace until age sixty.

*Pacesetters*

The final element of ergonomic adjustment is to look at work organization and pacing. The 'Overuse Injuries' fact sheet from the Canadian Auto Workers focuses its suggestions for preventive measures on work organization. Suggestions include training and education, as well as:

- slowing down the rate of work;
- abolition of the bonus system or piecework;
- no monitoring of work rate by machine or supervisor;
- adequate staff to cope with workload;
- rest breaks;

- no compulsory overtime;
- job rotation and/or task variation.[36]

Note that for assembly-line workers, poultry-plant workers, meat packers and other blue-collar jobs, the pace of work is almost always set by conveyor belts or computer-operated machinery. 'Over the past 15 years,' according to a story in the *Wall Street Journal*, 'line speeds in poultry plants have been revved up to a maximum allowable rate of 91 chickens a minute from the high 50s. (Industry and government didn't consider worker safety when ramping up speeds; rather processors convinced regulators that they could move chickens faster without sacrificing food hygiene.)'[37]

While some shops provide regular rest breaks, not all do. In meat processing, particularly, even a bathroom break may be problematic, because the company puts the bare minimum number of people to work on the line, and as much as a five-minute absence could hold up the whole line. Such workers have to ask their supervisor's permission to go, and their requests are monitored. In Alberta, the Alberta Federation of Labour had to intervene when one meat-packing plant began deducting time for bathroom breaks from workers' paycheques.[38] Such solutions as 'adequate staff to cope with workload' and 'no monitoring of work rate by machine or supervisor' may be impossible for a single worker to negotiate – they are generally feasible suggestions only in union shops.

---

*Three Rules for Preventing MSIs*

According to kinesiologist Dwayne van Eerd, the three top rules for avoiding MSIs are:

1. take frequent breaks;
2. take frequent breaks; and
3. take frequent breaks.

---

*Behind the Label*

In fact, the most important factor in prevention doesn't require any fancy equipment at all. Short breaks are relatively inexpensive, if not free. Conversely, much of what's sold as 'ergonomic' equipment is expensive and not necessarily proven. For instance, an ad in a recent issue of a computer magazine boasted about a new sleek tower that sits on the floor, not the desk, as offering a 'compact, ergonomic design.' Just how rounding the corners on a box makes it fit the worker better is not explained.

'Ergonomics' has become a marketing buzzword, and not only among those who buy and sell computers. Vegetable brushes, brooms, snow shovels, handbags – visit a 'healthy back' store on the World Wide Web and you'll find shelf after virtual shelf of products labelled ergonomic. Some are, some aren't. To compound the confusion, some products that have no ergonomic component to their design can nonetheless produce a temporary boost in productivity because of the 'Hawthorne [placebo] effect.'

Since trial and error can be a costly way to find out what works, the ideal method is to rent or borrow a product before actually purchasing it. Alternatively, in a large workplace, an employer can call in a consultant – often they're available at no charge or minimal charge from the local WCB or OHS agency – who can serve as a guide. CTD*News* noted that the first effect of California's ergonomic regulations may be finally to create a set of standards for ergonomics consultants.

THE ECONOMICS OF ERGONOMICS

Nobody's keeping it a secret: well-designed ergonomics programs actually save employers money. But many employers still have the perception that ergonomics programs must be fiercely expensive. High WCB costs – medical expenses, fines, or assessments – may be the prod that initially stimulates a company's interest enough to overcome this prejudice. That's generally when the company invites in an ergonomics consultant who eventually writes up the process for publication.

'The language of business is money,' said ergonomics consultant David C. Alexander. He noted that since 'reluctant compliance costs twice as much as foresight,' learning to talk money is 'a good way to leverage management into doing the right thing.'[39]

Gail Sater's video included pictures of a 'whisker trigger' stamping machine that required the worker to flick switches on both sides simultaneously – thus making it impossible for a hand to get caught and mangled. 'This machine cost $80,000,' she said, and put it into perspective by adding, 'That could be the cost of one Workers' Comp claim.'

Many of CTD*News*' 'best practices' examples[40] feature employers who are delighted when their lost-time days, WCB assessments, or medical insurance premiums drop significantly. For example, General Seating of Woodstock, Ontario, cut its lost workdays by seventy per cent in one year, from 1993 to 1994, mainly through job rotation, minimal line redesign, and involving workers in selecting new tools when old equipment was slated for replacement. As a bonus, the company discovered that keeping workers on the job – and off disability insurance – meant that job training costs dropped also.

Employers who look at their workplaces from a new perspective – as any good ergonomics program forces them to do – often find areas where their work practices can be streamlined for greater efficiency. In the same way that affirmative action reviews awaken awareness of hitherto unrecognized systemic barriers, so ergonomics reviews can identify production bottlenecks – not least because ergonomics reviews require direct employee participation. As Cindy Moser observed in an *OH&S Canada* editorial, 'employee health and company health go hand in hand.'[41]

When Cuddy Foods spent more than a million dollars to correct ergonomic problems, a noteworthy side effect was increased efficiency, which resulted in lower operating costs. *Canadian Business* quoted Cuddy's OHS coordinating officer Warren Beedle as telling his CEO, 'I'm going to give you back three dollars for every one you invest.' Did he succeed? 'It's ended up being six,' he said.[42]

Hank Lick said that Ford Motor Company did not achieve significant dollar savings by implementing ergonomics programs. However, addressing the company's health and safety problems was 'a jewel, a gem' that 'helped us in our relationship with the union,' he said. 'I think you look at ergonomics and totally what it does for you, your relations in your whole business ... Do you get back two dollars for every one? Maybe not. It saves money, but you're spending a lot of money to save it ... It's an intangible. A lot of people are in better jobs today because ... they're not hurting.'[43]

*OH&S Canada* reviewed six case studies of ergonomic adjustments and generally found that in the long run, the new set-ups not only decreased or eliminated injury costs; they also boosted productivity and efficiency. For instance, a make-up manufacturer changed the height of a conveyor belt and added new adjustable chairs with footrests – and increased productivity by 136 per cent. A cheque-processing facility changed workspace layout and job design and introduced two five-minute exercise periods and incentive pay – all at minimal cost – thereby drastically reducing overtime hours and lost-time claims due to work-related injuries. As well, faster processing meant cheques were deposited sooner, and the company earned more interest revenue.[44]

In short, when companies make the fateful decision to embark on ergonomics programs, the result is often less expensive and more rewarding than fretful forecasts from business media or business organizations may have led them to expect. An analogy may be drawn with environmental protection programs – the cost is often overestimated and the profitable side effects often overlooked.

In selling management on an ergonomics program, 'You can use one of three justifications,' concluded *OH&S Canada*. The first argument is that the company is legally or socially obliged to improve the quality of employees' working lives. The second argument is that ergonomics will reduce overhead and downtime due to absenteeism, compensation costs, turnover, and training time. The third justification is that improved ergonomics will increase production.

'The first justification concerning improved worker well-being is often not enough incentive for companies to implement a sound ergonomics program,' the magazine article summed up, somewhat ruefully. 'When the application of ergonomic principles results in net gain in profits for the company or industry involved, they are more likely to get the green light from management.'[45]

Obviously, the average worker is not likely to be in a position to gather enough information about her or his employer's finances in order to be able to make a persuasive cost-benefit analysis. And, equally obviously, companies that deny or attempt to refute employees' WCB claims are likely to resist an individual worker's suggestions for ergonomic improvements. For these and other reasons, when Barbara Silverstein was asked what an individual worker could do to protect herself in a workplace that was manifestly injuring other workers, she hesitated only a moment before responding with one word. 'Organize,' she said.[46]

# 9
# Beyond Grieving

It's an arresting image – an injured hand, wrapped in an elastic bandage, held out in supplication or in warning. Below it, in big bold type, the message: '*Last year more than 700,000 workers were hurt by carpal tunnel syndrome and other sprain and strain injuries. Newt Gingrich and other Republicans say we should do nothing about it.*'

It's the AFL–CIO's 1996 Ergonomic Action Alert flyer, ready to be downloaded from the Internet or photocopied hot off the fax machine, and which fanned out to union members across the United States.[1] It symbolizes the new assertiveness and political activism of a labour movement struggling to revitalize its purpose and rebuild a membership depleted by the dwindling away of smokestack and manufacturing industries – together with the workforce that ran them.

Even while employers' organizations mobilize to slash provincial and state Workers' Compensation budgets and to oppose implementation of ergonomics standards by local WCBs or OHS agencies, workers' groups are mobilizing to give legislators the workers' perspective. As the AFL–CIO flyer states, 'Crippling sprain and strain injuries like carpal tunnel syndrome are the #1 job safety problem in the workplace today.' Ergonomics is becoming a hot-button issue.

According to Peg Semanario, the AFL–CIO's Director of Health and Safety, the AFL–CIO is highlighting MSIs because 'we believe that the problem of ergonomics injuries and illnesses is indeed the major safety and health challenge that we face today ... The problem is simply too big, and too important, to ignore ergonomics; and indeed, too many workers are suffering.'[2] Not only are workers suffering, she went on to say, but some are permanently disabled in their thirties and forties.

Nor are unions as depleted or as powerless as some might believe. In Canada, the percentage of workers who are unionized has remained fairly constant. U.S. unions see opportunities in the late 1990s to rebuild their membership from the current fourteen or fifteen per cent of all employees, perhaps to the 1950s level of twenty-five per cent[3] – or perhaps more. Industries rise and fall, the workforce shifts from one sector to another – but workers still join unions.

Canadian unions have always had a stronger constituency than in the United States. 'In 1993, 32.6 per cent of workers were unionized (3,768,000). In 1966, 30.8 per cent (1,881,000) were union members,' reported the Alberta Federation of Labour (AFL). Drawing from a 1996 Statistics Canada report, AFL staffer Jim Selby noted that the seeming continuity 'masks a fairly dramatic shift in union membership as the manufacturing sector declines and the service sector grows. Male union membership has declined to 35 per cent of the workforce in 1993 from a high of 40.9 per cent in 1967. Women, on the other hand, have steadily increased their union membership from 18.7 per cent in 1967 to 29.8 per cent in 1993.'

WHAT COLOUR IS YOUR COLLAR?

Selby reports StatsCan's findings that union members are better educated, better paid, and more secure than non-union members. According to StatsCan:

- forty-five per cent of union members had a postsecondary certificate or university degree, compared with thirty-eight per cent of non-union members;
- the gap in hourly wages between union and non-union workers actually increased from $3.33 to $4.06 between 1984 and 1990 (in constant 1990 dollars);

- seventy-seven per cent of union members were covered by retirement plans in 1990, compared with only thirty-three per cent of non-union workers; and
- the average seniority of union members in 1990 was 8.8 years, as opposed to 5.5 years for non-union members.[4]

INFORMATION IS POWER

When Ramonalee K. first showed her supervisor at the poultry-packing plant that her right hand was swollen with tendinitis, the supervisor moved her to a spot where she used her left hand instead. Ramonalee worked until that arm swelled up, too. Then she left the plant, went to her doctors', and on the doctor's advice, filed a WCB claim. When the company disputed her time-lost claim and insisted that she return to work, Ramonalee turned to her union for information and support. By the time she presented her story to the 'RSI in the Workplace' conference less than a year later, Ramonalee was a shop steward, and other workers were coming to her for advice about their MSIs.[5]

Workers in the plant where Ramonalee worked belong to the United Food and Commercial Workers Local 373A, of which Vic Carr is president. The International UFCW provides Local 373A with superb information sheets about MSIs and other OHS hazards to distribute among its members. The sheets may be 'written in very simple language,' as Carr said, 'because so many of our members are new Canadians.' But they are excellent, concise documents, targeted at poultry plants and meat packers. Reading them, it's easy to see how the UFCW won public sympathy, rallied national boycotts of certain meat packers in the United States, and spurred the U.S. Occupational Safety and Health Administration (OSHA) to develop and enforce ergonomic standards for meat packing.[6]

*The Right to Know*

Informing workers about hazards is one of the original purposes of unions. As the Canadian Auto Workers' 'Statement of Principles: Health and Safety' puts it:

In our Canadian occupational health and safety laws, three rights are emphasized:

- the right to know about hazards of the workplace, especially chemical hazards;
- the right to participate in health and safety activities, especially joint worker–management health and safety committees; and
- the right to refuse hazardous work.

Workers demanded these rights through workplace struggles, strikes, and lobbying governments. We won these rights through the leadership of the New Democratic Party Government in the Province of Saskatchewan in 1972. Since then these rights have spread throughout the nation.[7]

LEST WE FORGET

One way that Canadian unions raise awareness of OHS issues is the National Day of Mourning, on April 28. 'This annual tribute to workers who were injured or killed in the workplace during the previous year,' according to New Brunswick Minister of Advanced Education and Labour Roly MacIntyre, 'was introduced by the Canadian Labour Congress on April 28, 1986, the anniversary of the day in 1914 when Canada's first workers' compensation legislation was introduced in Ontario. Last year, 17 New Brunswickers lost their lives in the course of doing their jobs, and workplace injuries and illness accounted for millions in compensation claims ... Sadly, workplace health and safety has not been a concept that has attracted a lot of attention from the general public – until tragedy strikes.'[8]

> *On the Toll*
>
> 'Each year more than 55,000 [U.S.] workers are killed as a result of occupational accidents or occupational diseases,' according to an AFL–CIO information sheet. 'Another 60,000 workers are permanently disabled, and nearly 7 million workers are injured on the job. That's one workplace death or injury every 5 seconds. Additionally, 25,000,000 workers each

year are exposed to toxic chemicals in their workplaces.'[9]

The Alberta Workers' Health Centre uses a similar statistic for Canada: 'Every seven seconds a Canadian worker is injured on the job. Someone is killed every six hours, and every day of the year, thousands of Canadian workers develop an occupational disease.'[10]

The Public Service Alliance of Canada stacks the stats another way: 'Every year some 1,000 workers get killed across Canada in workplace accidents which, in the majority of cases, are found to have been entirely preventable.'[11]

The twenty-eighth day of April is 'now officially recognized by Federal and Provincial governments, and many municipalities,' states a fact sheet written by Chemical, Energy and Paperworkers Union national representative on health, safety and the environment, Brian Kohler. 'By promoting education and understanding, the Day of Mourning can catalyze efforts to win improvements in workplace conditions.' The fact sheet suggests that on April 28, workers wear black armbands and ask to have flags flown at half-mast.[12]

The Public Service Alliance of Canada (PSAC) implemented a 'Spot the Hazard' sticker campaign for 28 April 1996. On that day, workers were asked to check their workplaces and notify their OHS representative or committee, who would slap stickers on hazardous material, machinery, or equipment.

'Because the workplace is dramatically changing and is becoming "leaner and meaner," the changes are having a detrimental effect on the health of workers,' PSAC warned. 'As a result, workers need to become more involved in protecting their own health.'[13]

In the United States, Workers' Memorial Day has moved to April 28, the date observed by the rest of the world. The AFL–CIO chose the 1997 Workers' Memorial Day to kick off its national 'Stop the Pain!' campaign for ergonomic awareness, distributing posters, fact sheets, a background booklet, and an Ergonomic Action Kit, along with an updated version of the 1996 flyer, which said: 'Every year more than 700,000 workers suffer from repetitive strain injuries. The National Association of Manufacturers, the American Trucking Association, UPS [United Parcel Service] and their Republican friends in Congress say OSHA should do nothing about it.'

Following on the success of the Day of Mourning, the Ontario Network of Injured Workers has declared June 1 to be Injured Workers' Day. In 1996 union WCB activists and injured workers' groups organized demonstrations and protests on June 1 against the Tories' proposed WCB changes.

BY THE BOOK – ER, THAT IS, THE MANUAL

Ergonomics represents a 'major problem' for PSAC members, who are mostly office workers, said health and safety officer Frank Hamade.[14] Although there are no federal ergonomics standards (or a federal OHS enforcement agency, for that matter), the Canadian Standards Association and the Treasury Board Manual do provide guidelines. The Manual considers ergonomics in some detail: 'Managers should think "ergonomics" when replacing equipment, setting up a new workplace, modifying tasks, or planning changes to the old ones ... Ergonomic considerations in the early design stages of the VDT workstation as well as the tasks to be accomplished will minimize safety and health hazards as well as maximize employee productivity.'[15] Even with precise ergonomics prescriptions, cutbacks in the public service and increasing workloads on remaining workers have meant that PSAC members are still at risk for MSIs.

SHADES OF *NORMA RAE*

Conventional wisdom has it that workers don't go on strike about health and safety issues. Wages, benefits, working hours – those are supposedly the key factors in worker (dis)satisfaction. Cathy Walker doesn't agree. 'I think that health and safety issues in a number of situations have been strike issues,' she said, 'but you tend not to see that kind of thing reported, at least not *as* health and safety issues.'[16]

Other concerns that are really OHS issues include hours of work, mandatory overtime, and contracting out. In each case, the question is job design – ergonomics. For instance, major media reported that the Canadian Auto Workers union's CAW negotiations with the Big Three auto makers centred on contracting out (which threatens the core workforce) and mandatory overtime (working with fatigued muscles). A key victory for the United Auto Workers was the appointment of three new UAW national ergonomics coordinators, one each for Ford, General Motors, and Chrysler.

MSIs also surfaced as strike issues in themselves in the two big Alberta strikes during the first half of 1997. Safeway workers ran a series of newspaper ads during their seventy-five-day strike, including one that said, about a cashier's job: 'On a typical shift, I bend, twist, pull and rotate thousands of pounds of groceries through my checkstand. Some days my arms hurt so much, I can't lift my child when I get home. My doctor says I may have a repetitive strain injury.'[17]

And when a *Calgary Herald* reporter talked to striking workers outside the Cargill meat-packing plant, those workers talked about MSIs too. 'Many workers crowded around a *Herald* reporter at their picket line Saturday, eager to show their heavily calloused hands and describe the pain of repetitive strain injuries ... [Roger] Gill, who helps train new employees, said the assembly-line jobs require workers to repeat the same two or three motions again and again, while handling heavy slabs of beef. The inevitable result is tendinitis, strained muscles and other ailments ...'[18]

THE PARTIES OF THE FIRST PART ...

'During the mid- to late 1970s,' according to Vernon Mogensen, 'unions began to press employers for collective bargaining language that included ... ergonomically designed furniture and equipment' for computer users.[19] Some unions have succeeded, especially in the professions. In the United States, the 1996–7 Wire Service Guild agreement with the Associated Press has specific clauses on workplace noise and rest breaks, in addition to RSI-specific language incorporated in the 1992–4 Health and Safety Article.[20]

In Canada, the Southern Ontario Newspaper Guild (SONG) contract with the *Globe and Mail* provides up to three thousand dollars in extra medical expenses for any staff member who suffers an MSI – a clause brought about largely through the efforts of reporters Joan Breckenridge and Paul Taylor, whose careers have been adversely affected by keyboard injuries (and voice-recognition software injuries, too). RSI coverage was extended in the 1996 agreement. In March 1996, SONG launched *RSI Watch* – one of the largest workplace studies of its kind in Canada – in collaboration with the *Toronto Star* and the Institute of Work and Health.[21]

At the Casino de Montréal, workers announced in June 1996 that, after a year-long strike, they had won a four-day work week. Following thirty-four arbitration sessions, a dozen hearings and two visits to the Casino to view the conditions under which the 1,400 union members work, the arbitrator imposed a two-year collective agreement on workers and management. Among workers' concerns (along with job security) were backache, tendinitis, and burnout: 'Le casino étant un milieu de travail présentant des conditions propices, entre autres, aux maux de dos, aux tendinites ou au *burn out*, toutes les questions touchant la santé et la securité au travail revêtaient une importance majeure.'[22]

*On the Dash(board)*

Unions based in the automotive industry have taken a lead in dealing with MSIs and ergonomics issues – just as some auto manufacturers have done. The United Auto Workers in the United States and the Canadian Auto Workers (CAW) have not only developed significant expertise in auto plants, but they've cultivated ergonomic skills among workers in all the other workplaces where they have members.

MSIs 'are just about our biggest health and safety issue today,' said Cathy Walker, Director of CAW's Health and Safety Department. 'In fact, it's probably the one unifying health and safety issue in our union, just because everybody has to deal with it. It doesn't matter if you're working in a restaurant, dealing with tendinitis because the plates are too heavy, or if you're a miner working underground.' Experience with MSIs in Windsor auto plants translates directly into policy guidelines for MSI prevention in Windsor casinos.

*Rainbow Collars*

Anyone who is puzzled by the idea that CAW members might be waiting tables hasn't been paying attention to recent organizing drives. Unions have been diversifying, expanding into new industries, and trying to serve the changing needs of the changing workforce. CAW's 205,000 members work in twelve different economic sectors, including airlines, fisheries, surface transportation, hospitality, and general services. About a fifth (40,000) of the CAW's members are women.

Or let's look at the Office and Professional Employees' International Union, which celebrated its fiftieth anniversary in 1995. 'We have become a vibrant organization,' boasts OPEIU's home page, 'representing "white collar" professionals such as computer analysts and programmers, data entry operators, copywriters, nurses, health care certified and licensed employees, doctors, attorneys, hypnotherapists, models, artists, museum curators, law enforcement officers and security guards, accountants, engineers, secretaries, bank employees, insurance workers and agents, Wall Street employees, and many, many more classifications.'[23] Similarly, the United Steelworkers of America not only boasts members right across Canada, but also represents 5,000 workers in Retail–Wholesale Canada, in workplaces ranging from department stores to fast-food outlets.[24]

*Ergonomic Outreach*

In the face of resistance from individual employers, unions have sought other ways to educate, inform, and protect their own members – as well as the rest of the workforce. In 1992 the Women and Work Research and Education Society of Vancouver held a conference on MSIs and produced a sixty-four-page booklet for BC workers titled *Repetitive Strain Injuries in the Workplace*.

Women and Work coordinators Lois Weininger and Lynn Buerkert then expanded the project and repeated the process in Alberta and eventually in Nova Scotia.

Another approach is represented by Workers' Health Centres, which have been established in Manitoba, Ontario, Alberta, and PEI, where anybody can call for information and referral on OHS issues. These centres also provide peer counselling, lists of WCB worker-advocates, and OHS courses for workers and small-business owners.[25] The Edmonton centre acquired audiometric equipment so that it could offer hearing tests.

Ontario workers with work-related injuries can find help at the Occupational Health Clinics for Ontario Workers (OHCOWs) in Toronto, Windsor, Sudbury, and Hamilton. Funded by the Ontario WCB and sponsored by the Ontario Federation of Labour, the OHCOWs do not treat individual patients – their purpose is prevention – but they do diagnose ailments and then educate the patient's family doctor on how to deal with occupational diseases. Whether union or non-union, public or private sector or self-employed, patients go to OHCOW because it's a place where they can find healthcare providers who already understand conditions such as MSI and can analyse the work situations that cause them. The OFL also oversees the Workers' Health and Safety Centres, funded by the Ontario WCB, where workers can get OHS training, including training in ergonomics.

*Who's Afraid of the AFL–CIO?*

At the end of 1995, John J. Sweeney became the new AFL–CIO president, with Linda Chavez-Thompson as vice-president. The union suddenly kicked into high gear, recruiting high-profile organizers from other unions and from the federal government, organizing Town Hall meetings from coast to coast, funding a new twenty-million-dollar Organizing Department, revitalizing central labour councils, and unleashing attack ads against conservatives in Congress.

'Labor's muscle-flexing has changed the political equation,' the *New York Times* reported in May 1996. 'By plunging the labor movement into politics as never before, Mr Sweeney has got business lobbyists and political conservatives to snap to attention.'[26] The article credits union lobbying as instrumental in getting the federal minimum wage raised.

'Expectations for Union Summer were surpassed,' crowed the AFL–CIO 1996 Labor Day media release. 'More than 3,000 people from 45 states applied for the internships.' Some 1,500 students were selected, paid nominal stipends, and assigned to raise the union profile, help workers organize, and lobby for worker-friendly legislation. 'In their first two months,' the union reported, 'the interns distributed more than 150,000 leaflets, made about 5,000 visits to workers in their homes, visited 1,500 work sites, helped organized 235 protests and rallies and organized workers who speak 10 languages, from Spanish to Tagalog.'

In May, the *New York Times* reported a public opinion poll showing that Americans sided with unions over management 41 per cent to 24 per cent – more than double the margin of approval of three years earlier. In July, *Newsweek* reported that 62 per cent of Americans had a positive view of unions.[27]

WOMEN AND UNIONS

As momentum builds, as unions gather renewed strength, it's easy to forget that traditionally unions have been concentrated in heavy industries with membership comprising mainly males. 'Like employers, many labor leaders viewed working women as being marginal to the labor force,' noted Vernon Mogensen.[28] Perhaps only a cynic (or a feminist) might ask wistfully which unions really represent the women in retail, clerical, electronics, meat packing, poultry packing, postal work, medical transcription, data entry, hairdressing – the workers who are most at risk for work-related MSIs?

---

*Plus Ça Change ...*

'The low wages at which women will work form the chief reason for employing them at all ... A woman's cheapness is,

> so to speak, her greatest economic asset. She can be used to keep down the cost of production where she is regularly employed. Where she has not been previously employed she can be introduced as a strike breaker to take the place of men seeking higher wages, or the threat of introducing her may be used to avert a strike. But the moment she organizes a union and seeks by organization to secure better wages she diminishes or destroys what is to the employer her chief value.'
>
> — U.S. Bureau of Labor, *Report on Conditions of Women and Child Wage-Earners in the United States*, 1911.[29]

*Plus C'est la Même Chose?*

'While more and more women are working,' said Lin Lim, author of a new International Labour Organization (ILO) report, 'the great majority of them are simply swelling the ranks of the working poor. Women's economic activities remain highly concentrated in low-wage, low-productivity and precarious forms of employment.'[30]

With a title (*More and Better Jobs for Women*) that's more a call to arms than a summary, the ILO's July 1996 global study offers a few sparkling highlights amidst the usually bleak employment landscape. Women have made progress in professional and managerial jobs, in reconciling workplace and family responsibilities, in joining trade unions and sensitizing them to gender issues, and in achieving pay equity laws in both developed and developing countries.

However, ILO Director-General Michel Hansenne stated, 'Equality of opportunity and treatment for women in employment has yet to be achieved anywhere in the world.' Overall, women in both developed and developing countries work longer hours than men and are paid an average of twenty-five per cent less.[31]

Indeed, the ILO always has been visionary – its first ergonomics convention passed in 1977 (Working Environment, Convention #148) and its second in 1981 (Occupational Safety and Health

Convention, #155). Nations that signed these conventions have agreed to research, control, monitor, report statistics annually, educate workers, and hold inquiries into occupational accidents, occupational diseases, and other injuries to health that arise in the course of or in connection with work. All local unions have to do is hold their governments to their agreement.

Apart from ILO reports, women workers in Canada and the United States – especially in non-unionized workplaces – might well wonder what local unions have to offer them, or whether local unions are interested in them at all. More, the very thought of taking on yet another activity – trying to juggle union meetings in addition to family obligations – may put women off the idea. Finally, unions have earned a blanket reputation for treating women poorly.

*Up against It*

'The labour movement was deeply ambivalent about women, to put it mildly,' wrote author Susan Crean, reviewing labour history of the 1930s and 1940s in her biography, *Grace Hartman: A Woman for Her Time*. 'In both world wars it supported the idea, but only as a last resort; that is, women could work [outside the home] if there were no men to be found anywhere first ... labour's initial support for equal pay often seemed to have more to do with protecting men's jobs and job standards than improving wages for women. In fact, in some instances, unionization worked to the immediate disadvantage of women, as employers, when forced to raise wages, refused to pay the higher rate to women and hired men instead.'[32]

Whether in or out of the paid workforce, women were deterred from organizing because of their family roles. Quebec union organizer Madeleine Parent spent the 1940s among Quebec textile workers, of whom about 40 per cent were women, with some as young as ten years old. Women working in the textile mills were especially concerned about what was then known as 'favouritism.' 'We didn't have the term sexual harassment in those days,' recalled Madeleine Parent, 'so we called it favouritism.'[33]

In order to organize women workers, in a time and place that penalized women who attended meetings by themselves, the union invited women workers to bring their families to meetings. The women pulled up their chairs and formed an inner circle, where they conducted their business. Their husbands (whose own jobs were elsewhere) and their children sat in an outer circle, around the perimeter of the meeting room, and the children ran back and forth between their parents. This arrangement allayed the men's concerns and also solved the problem of childcare – only one of the external deterrents to women participating in unions.

'Almost two-thirds of the workforce remain without a union, and most of those workers are women,' union organizer Laurell Ritchie wrote in the 1983 book, *Union Sisters*. Ritchie suggested several factors that hamper organizing efforts among women workers. Women tend to work in smaller workplaces, where they have less bargaining power. New immigrants are often segregated by employers and set to work in different departments, which makes it harder for them to organize across the company. One-quarter of women in the workforce (at that time) worked part-time, and Labour Department regulations often excluded part-timers from union eligibility.

Finally, Ritchie addressed the kinds of jobs that women hold: 'Three-quarters of the female labour force work in the "newer" industries of trade, finance, public administration, business and personal service – areas that do not have a long history of organization.'[34] Aside from public administration, these 'newer' industries have grown considerably since 1983. Throw in computers, and you'd have a good list of where most of the job growth has occurred in the last decade.

*One Step Backwards, Two Steps Forward?*

Women have been most likely to be unionized in the public sector, such as administration, where Grace Hartman worked – which of course has been pared back sharply in the 1990s. When the province of Alberta privatized liquor stores, for example, the

Alberta Union of Public Employees (AUPE) lost 1,500 members at one blow. After two years of Ralph Klein's government, AUPE's membership was halved, from 38,000 to 19,000. But as Linda Karpowich, (then) president of the Alberta Federation of Labour (AFL), remarked in a 1994 interview, at the same time the AFL's membership in the public sector dropped, its membership among private sector employees rose. And since the AFL accepts part-time workers as members, for some sectors (such as liquor stores), the change resulted in a net increase in AFL members.

And, as Jim Selby mentioned in the AFL's *Labour News*, the proportion of women who are union members has climbed in the last thirty years. This new and rather surprising increase did not develop accidentally. Even as Laurell Ritchie enumerated the barriers to recruiting women, she and her union sisters were mobilizing to overcome those barriers.

In the 1990s, according to Canadian Labour Congress executive vice-president Nancy Riche, women joined unions eight times as fast as men.[35] One reason is that unions have been supporting local women's groups in events such as the annual Take Back the Night marches and International Women's Day. Certainly the 1996 cross-Canada Women's March Against Poverty would not have been possible without the support of the CLC and local union affiliates.

Similarly, U.S. unions have been instrumental in mounting opposition to the neoconservative agenda. In California, labour has supported and spearheaded the public campaign to save affirmative action programs, threatened by a referendum ballot, and participated enthusiastically in Fight the Right rallies.

As well, unions have adopted some of the issues put forward by women's groups, and recognized their impact in the workplace. Childcare has been incorporated into collective agreements. Domestic violence is now an issue that unions address among male and female workers. Sexual harassment, pay equity, part-time work, underlit parking lots where sexual assailants might loiter – canny unions now treat such issues as mainstream, not marginal.

In the public sector, the Public Service Alliance of Canada has issued a booklet called *A Level Playing Field*, describing its fifteen-year campaign to implement pay equity (equal pay for work of comparable worth). 'Starting with our female-dominated bargaining units in the federal public service,' wrote Louise Laporte of PSAC's National Action Program, 'we have expanded to other bargaining units under the Canada Labour Code, and provincial and territorial legislation ... We need to talk with other unions and labour bodies and our coalition partners to ask them to assist us in pressuring governments and employers to meet their pay equity obligations.'[36] PSAC also sponsors regional and national women's conventions, recently hosted its first Native and ethnic minority convention, and has fought for (and won) recognition for same-sex family benefits, such as bereavement leave and extended healthcare insurance.

Then there's the Canadian Union of Public Employees. 'When Grace [Hartman] was elected president of CUPE in 1975, it wasn't a *glass* ceiling she had to break through. The ceiling was made of iron,' reflected Judy Darcy in an afterword to Susan Crean's biography of Hartman. 'When I was elected president and Geraldine McGuire secretary treasurer in 1991, the fact that it was barely an issue that the two top jobs in the union would be held by women, spoke volumes about how far our union had come.'[37]

National unions now are reaching out to women, asking about their working conditions and inviting them to participate in change. 'For women, full-time, full-year jobs have virtually disappeared. In 1996, all of the growth in jobs for women was in the part-time category,' warned the authors of *Women's Work: A Report by the Canadian Labour Congress*. Project coordinator Winnie Ng collected interviews with women across Canada whose voices enliven the grim statistics she dug up. Half the book described the problems women face in both paid and unpaid work; the second half outlined social action towards solutions.[38]

In the United States, the AFL–CIO has launched its campaign, 'Ask a Working Woman,' which includes a national survey and a nationwide conference (September 1997), in order to develop a

'Working Woman's Agenda, to present to the President, the Congress, the National Chamber of Commerce, and the media.'

PUBLIC STEWARDS

Union protests about public sector job losses have coincided with some degree of public apprehension about loss of public services. Maybe the Alberta public didn't mind when liquor stores were privatized. But Canadians everywhere are very worried about the quality of healthcare services, and somewhat worried about public education. Canadian unions have tried to address that anxiety and regain relevance in the public mind through coalitions such as Action Canada and Alberta's Common Front. Union-led demonstrations in Alberta and Ontario raised union profiles in the minds of a public who have grown sceptical of claims that unfettered free enterprise will provide prosperity and (cheap) services for all.

In the United States, the AFL–CIO has kicked off its public 'Stand UP for America's Working Families' campaign. Subtitled 'The men and women of the AFL–CIO leading the fight for America's working families,' the local campaigns are targeted at such issues as 'good jobs' and worker protection. 'More workers today are employed by fast food restaurants than in the steel and auto industries combined,' according to the union's analysis, 'and the largest single employer in the U.S. is Manpower Inc., a temporary help agency, which now employs twice as many people as General Motors.'[39]

*Who You Gonna Call?*

As high unemployment persists and workers deal with job pressures and insecurities, unease with laissez-faire capitalism may make members of the public more receptive to the union message. When the news that unemployment has fallen causes the stock market to drop like a brick, as happened in the spring of 1996, many people get the perception that the basic social contract has been broken.

When companies demand that ordinary workers take wage rollbacks while top executives award themselves six-figure bonuses, and dedicated long-term workers are laid off permanently en masse because of company takeovers, workers start to question their loyalties to their employers. When job loss means dependence first on reduced Employment Insurance and then on reduced welfare – both with rigid, seemingly arbitrary benefits limitations – people start to feel desperate about their lack of options.

---

*People vs. Corporations*

'A telephone survey of 800 voters during June 1996, accompanied by six focus groups, has revealed that Americans of all ages, all incomes, all races, and both political parties are seething over the way large corporations have been treating their workers in recent years. Specifically, voters are angry about the following kinds of behavior:

- Massive layoffs at a time when profits and CEO salaries have been going through the roof;
- Firing full-time workers and filling their jobs with temps who get neither medical benefits nor pensions;
- Firing older workers and replacing them with younger workers who are satified with lower wages.

In sum, people are angry that corporations are showing disrespect for, and disloyalty to, the American workforce.'

— *Rachel's Environment & Health Weekly*, no. 507, 15 August 1996.[40]

---

IF THEY DIDN'T EXIST ...

The 1990s have seen union-busting campaigns in both public and private sectors. In response, unions are metamorphosing, so that when (say) public service unions lose members to cutbacks, private sector unions pick up those members in their new jobs. In fact, union memberships seem to be resurging in the mid-1990s, after a slump.

But there's also an interesting trend towards new organizations trying to reinvent ways to do what unions do – or to shore up workers who don't have access to unions. Take the Interfaith Committee for Worker Justice in North Carolina, where local churches undertook to support poultry-plant workers' demands for better working conditions. Unions face tremendous prejudice trying to organize in the South; church-based groups such as the Interfaith Committee for Worker Justice command more respect locally.[41]

'The conditions of work are the most important thing for us in this strike,' a picketer told L. Paltrineri and Don Mackle, reporters for the *Militant*. According to the reporters: 'A delegation of religious leaders visited the plant in April, and documented many of the unsafe conditions workers face. Some of the problems include management's refusal to allow workers to go to the bathroom between breaks, and high line speed resulting in many injuries, from carpal tunnel syndrome to knife wounds.'[42]

Perhaps less altruistic is this approach: 'From the 1950s to the present, labour organizations have not changed either their goals or their structure,' claims a brochure from the fledgling National Congress of Employees (NCE). 'Employees today, especially the approximately 85 million that have remained unorganized, have found themselves in a vacuum of uncertainty. As workers of the 1930s experienced an inarticulate frustration that found its dynamic expression in the American Labor Movement, employees today are experiencing an inarticulate dissatisfaction in the employment condition ...' NCE's proposals tend to be apolitical, such as discount buying services.[43]

Another new organization – this one with some links to established labour unions – is Workers of America. The group's 'Summary of Major Worker Issues' includes wages (specifically, minimum wage), benefits, job security, training and education, corporate responsibility, taxes, foreign competition, the workplace environment, national industrial policy, infrastructure, and political reform.[44]

At *Disgruntled* e-zine (http://www.disgruntled.com), which publishes stories about the horrors of modern working life, some readers have joined together to form the *National Employee Rights Institute* (NERI), with close links to the National Employment Lawyers' Association. NERI includes an information clearinghouse on employment rights, a law school clinic project, legislative advocacy, and a think tank.[45]

Whether these efforts actually attract significant membership is almost beside the point. What they share is an expression of the public's growing distrust of the employer–employee relationship.

Other, similar groups and initiatives include the following:[46]

- Working Today coordinates health insurance for consultants, freelancers, employees, temps, and self-employed people as part of its mandate for advocacy, education, and service. It also hosts a national Good Jobs campaign, calling for portable benefits;
- Jobs With Justice is a national labour, community, and religious coalition dedicated to fighting for the rights of working people, with two dozen local coalitions in the United States;
- Jobs For All has called for a national job vacancy survey to demonstrate that unemployment is structural and workfare requirements are unenforceable;
- Texas-based Americans with Work-Related Injuries conducts public education on work-related injuries and their employment consequences;
- America's MedCheck Organization calls for national Workers' Compensation standards and unhindered access to quality medical care for injured workers, along with guaranteed return to work;
- Amy Clipp's book, *Job Damaged People*, lists twenty-nine Committees on Safety and Health across the United States and Canada; these conduct research and lobby for improved job safety and health legislation.

Wages, hours, and benefits are the job issues workers usually name when someone asks what matters most to them. When workers can find new jobs that offer better wages, hours, or benefits, they are likely to quit their old jobs and move on.

Workplace hazards are another matter altogether. 'Contrary to the traditional theory of worker ignorance and indifference, [this] study finds that a very high percentage of workers do recognize major hazards on the job,' according to a meta-analysis by James C. Robinson of the University of California. He continued, 'In contrast to the existing economic studies of worker responses to hazard information, the analysis finds that collective 'voice' responses [speaking out] were more common than individual "exit" responses [quitting] during the time period under consideration.'

Drawing on data from five separate studies, Robinson concluded that, 'hazardous working conditions have been found to be associated with higher levels of expressed dissatisfaction, absenteeism, discharges for cause, authorized and unauthorized strikes, and (though not consistently) higher quit rates.'[47]

*Solidarity – One Way or Another*

Workers seem drawn to forming labour organizations. Adversity only makes such organizations more attractive – in the developing world as well as in industrialized nations.

'Fifty years ago,' Canadian Auto Workers president Buzz Hargrove reflected on Labour Day 1996,

Ford workers in Canada made just under a dollar an hour and struck over the right to have a union. In Mexico, Ford workers struck in the 1990s over the right to have a democratic union ...

Unions were formed out of the struggles of workers to improve their conditions and to bring justice and dignity to their lives, their families and the communities they live in ... [Now] corporations and right wing governments are gleefully pushing low wage, anti-worker policies. Wherever possible they're changing laws to reduce the freedom of workers to join unions and the freedom to come together to achieve change. They fear labour because it is one of the few institutions left that can effectively challenge increasing corporate power. They have worked hard to undermine unions by labelling them as creatures of the past. Nothing could be farther from truth. In fact it is corporations that want to recreate the past by increasing their global size and power and ignoring borders as they move work to the lowest wage economy ...'[48]

Of course, corporations and business interests rebuff such charges with assertions that they are just trying to preserve the nation's economy in the face of globalization. Consider the *Globe and Mail*'s 1996 front-page report on rising WCB costs. After discussing the increase in MSIs and the putative causes, reporter Susan Bourette added, 'Still, the real issue for business is learning how to reduce the exorbitant costs that bite deeply into the ability of Canadian companies to compete globally.'[49]

IN THE NAME OF COMPETITION

Aside from the pat rejoinder that the obvious way to reduce costs is to prevent injuries, there are two other arguments against discussing WCB costs in the context of global competition. First of all, workers in other countries are just as susceptible to injury as Canadians or Americans are. And as Buzz Hargrove mentioned, Mexican workers are forming unions, too. So are South and Central Americans, in response to the North American Free Trade Agreement (NAFTA), and Asian, Indonesian, and Phillippine workers, in response to the Asia Pacific Economic Community (APEC).

For instance, delegates from trade unions in Colombia, Ecuador, Peru, and Venezuela met in Caracas in August 1996 to discuss democracy, development, and the status of the labour movement. As a result of that meeting the Regional Inter-American Organization of Workers (ORIT) set the theme for its upcoming fourteenth convention – planned for April 1997 in Santo Domingo – as 'The Alternative: Democratizing Globalization.' According to the InterPress Service report, 'the Caracas conference identified a series of common problems, such as unemployment, the loss of purchasing power of wages, and the growth of the informal sector. Women are especially vulnerable in this sector, which lacks any kind of social security.'[50]

Taking the global view, the Canadian Labour Congress finds that 'in 1996, the world economy is growing at less than half the 1973 rates, long-term unemployment has risen dramatically and, in the vast majority of countries, wages have been virtually frozen at the level of the early 1970s while basic labour standards have been seriously eroded.' As a way of working towards the return of responsible governments, the CLC has pledged 'to promote and build coalitions and social movements around the world to advance common interests.'[51]

This brings us to the second, perhaps more powerful point in this discussion. When the business community claims that enforcing OHS standards would hamper U.S. and Canadian industries in their competition with other nations, what they are really announcing is that they have already made a policy decision that the richest nations in the world will compete against the poorest nations of the world. Because if the United States and Canada decided that their main economic competition was with other developed nations, then they would have to recognize that most developed nations – particularly those of the European Union and Japan – have had ergonomics legislation in place for years.

# 10
# Legislation in Other Jurisdictions

IT'S A CRIME

'Failure to comply with the regulations could put you in the dock – either in the Magistrates' Court or, if the Inspectorate regards the offence as sufficiently serious, in the Crown [criminal] Court.' That's how *Accountancy* magazine braced British professionals to deal with the new (in 1992) European Community directive on Visual Display Units.[1]

While emphasizing environmental factors such as temperature, ventilation, and lighting, the EC legislation 'also focuses on ergonomic factors,' the article advised. 'For instance, VDU users will have to be provided with adjustable chairs and foot rests.'

For *Accountancy* readers, the kicker is that British law takes occupational health and safety very seriously. Inspectors make random spot checks of workplaces and have the option to lay criminal charges. 'The outcome of recent cases makes it perfectly clear that the law does not take claims lightly. Individuals with RSI have already been awarded as much as £45,000 in compensation, and there are unlimited fines for breaking the health and safety laws,' warned the article.

Or, as the U.K. Health and Safety Commission prefaced its 1992 update of the Approved Code of Practice, 'Although failure to comply with any provision of this Code is not in itself an offence, that failure may be taken by a Court in Criminal proceedings as proof that a person has contravened the regulation or sections of the 1974 Act to which the provision relates.'

The Code – the equivalent of the Occupational Safety and Health Administration's (OSHA's) general duty clause – sets out the minimum standards for management of health and safety at work, and requires written records of risk assessment and steps taken for employee protection. Ergonomics is implied in such clauses such as 27(c), *'wherever possible, adapt work to the individual* especially as regards the design of workplaces, the choice of work equipment and the choice of working and production methods, with a view in particular to alleviating monotonous work and work at a predetermined work rate.'[2]

*Ergo Spelled Out*

Ergonomics was central to the 1994 British Health and Safety Executive (HSE) National Health and Safety in the Workplace Week campaign on MSI prevention. The HSE announced that more than 2,000 companies had signed a pledge to take action to prevent MSIs at their worksites. This was part of an ongoing campaign called 'Lighten Your Load,' which was initiated in 1991, and featured a new publication titled, 'A Pain in Your Workplace?' as the latest in an ongoing series of information books and pamphlets. The chair of the HSE Commission, Frank Davies, estimated that there were up to 110,000 cases of upper-limb disorders in England and Wales in 1990. 'No industry is exempt from these problems,' he said in a media release. 'HSE estimates the loss of 5.5 million working days per year, and a cost to employers of £1.25 billion.'[3]

While the HSE enlists employers in meeting European standards, the Trade Union Congress, a labour umbrella organization, sets about educating workers. The Congress's 'Don't Suffer in Silence' campaign involves collective bargaining, posters, publications, training to recognize hazards and symptoms, and a national conference. Among other resources, the Trade Union Congress offers a leaflet on *Women and WRULDS* (work-related upper-limb disorders) and a poster titled 'Women at Risk.'[4]

Despite all these efforts, the reported MSI rate continues to rise. The Trade Union Congress cites a British Safety Council report showing a 'staggering increase' in the number of people compensated for carpal tunnel syndrome: 267 cases in 1993–4, up from only twenty cases in 1992–3. The BSC has joined the struggle by launching a new publication, *Safety Attitude Guide: Preventing RSI*.

---

*Does Your Other Hand Know?*

Sometimes governments seem to set up different policies to work at cross purposes. In the United Kingdom in 1996, even as the Health and Safety Executive, Trade Union Congress, and British Safety Council were campaigning to raise awareness of MSIs and get workers to pay attention to early symptoms, the government also cracked down on eligibility for Workers' Compensation – with a predictable result:

'The union representing Benefits Agency doctors, the IPMS, has revealed that the extra workload of medical tests under the new Incapacity Benefit has caused an outbreak of writer's cramp. Many BA doctors are themselves examining RSI sufferers desperate to keep their benefit rights – the cause of the problem is the 28 assessment form, introduced to make it more difficult for unemployed people to claim the benefit.

'IPMS negotiator David Luxton commented: "There is a tremendous backlog of casework, directly attributable to over-ambitious Government targets. Many doctors have complained that they are suffering from writers' cramp symptoms, or tenosynovitis."'[5]

---

Note that while British efforts to control MSIs have not yet been entirely successful, at least the government has acknowledged that such injuries exist and has published some real statistics on the incidence rate. Moreover, the picture that emerges suggests the whole country is concerned about the problem. There are problems of adjustment, to be sure. The Trade Union Congress reported that while British Telecom and the Communications Workers Union have negotiated an agreement that video coders should have short and frequent rests of ten minutes out of every seventy – in line with HSE advice – a membership survey showed that these rest breaks were the ones employers are most likely to pressure staff to give up. Still, the government guidelines alone offer more support than most North American data-entry workers have – much less the negotiated agreement.

HAND IN HAND

The United Kingdom's VDU standards owe much to the European Union – literally, in the case of the Trade Union Congress's campaign briefing newsletter, which states that it is funded by the European Commission. Contrary to NAFTA (which tends to treat labour standards as barriers to trade), in the European Union, compliance with EU labour standards is a requirement for participating in the free market.

'Standards are necessary to provide quality control and to support legislation and regulations used in the establishment of an equal opportunity and fairly operating international market,' explains British ergonomist K.C. Parsons in an overview paper published in *Applied Ergonomics*. 'European standards are produced (replacing national standards) and the following countries are bound to implement them: Austria, Belgium, Denmark, Finland, France, Germany, Greece, Iceland, Ireland, Italy, Luxembourg, Netherlands, Norway, Portugal, Spain, Sweden, Switzerland and the United Kingdom.'[6]

Workplace ergonomics was chosen as one of several key concerns in 1987, when the *Single European Act* included a specific clause on occupational health and safety, known as Article 118a. Although European states had been cooperating on health and safety issues since the 1957 agreement to monitor coal and steel workers, Article 118a introduced a new impetus.

'There is widespread public support for Community action in this field,' according to a report from EU Occupational Health and Safety. 'For example, a Eurobarometer survey carried out during the European Year of Safety, Hygiene and Health Protection at Work ... revealed that a large majority of the Community's working population were in favour of introducing minimum common requirements at Community level ...'[7]

*Par Example, en Français ...*

As with Britain, France has legislation that requires employers to provide computer workers with footrests, on request. The FranceWeb site boasts a series of pages on 'the risks of working at a computer' [les risques du travail sur ordinateur]: 'A number of scientific studies have shown that, when we ignore (don't respect) certain ergonomic constraints, computer work can become *dangerous to our health*. This danger increases considerably when we spend more than four hours a day in front of our screen.' [De nombreuses études scientifiques ont démontré que, quand nous ne respectons pas certaines contraintes ergonomiques, l'ordinateur peut devenir *dangereux pour notre santé*. Ce danger augmente considerablement quand nous passons plus de quatre heures par jour devant notre écran.]

A lengthy section on wrists discusses forearm position at keyboard, keyboard types and keyboard trays, and total keyboarding hours. It recommends a five-minute break every hour or a fifteen-minute break every two hours, and urges anybody who has the slightest concern about his or her wrists to see a specialist promptly, to prevent permanent damage.

One clever suggestion in this section is a 'pencil test' to check for wrists in neutral position: 'Place a pencil lengthwise on your wrist, from the base of your fingers. Work normally at your keyboard. Does the pencil stay in constant contact with your skin? If so, you're maintaining a good position.' [Posez un crayon, en longueur, de votre poignet au début des articulations des doigts. Travaillez normalement sur votre clavier. Le crayon reste-t-il en contact permanent avec votre peau? Si oui, vous tenez la bonne position.]

French law states that computer workers must have tiltable keyboards that are separate from the screen, with enough space for a wrist rest or forearm rest, and that the keyboard must have a reasonable touch.

The page about back pain states explicitly that, if you sit poorly at your screen, it's the workstation that is causing your problem. [Si vous vous tenez mal devant l'écran, c'est que votre poste de travail est mal adapté à votre posture – and non pas le contraire.] French law also requires an adjustable screen with soft lighting and enough contrast that characters can be read from a normal working position.[8]

*In Front by a Decade ...*

Although the EU strives to achieve minimum occupational health standards among member countries, that doesn't necessarily mean uniformity. Some countries are farther along than others. Scandinavian countries, for example, have been studying MSIs for at least two decades, and have established solid research on incidence and occupational links.

'Musculoskeletal disorders are a common health problem,' according to Professors Mats Hagberg and Asa Kilbom of the Swedish National Institute of Occupational Health. 'During the last year about four of ten Swedes have had back problems. Aching necks and shoulders affect 15 to 20 percent of the population every year ... In Sweden, one woman in ten and one man in 15 is handicapped by [MSIs] to some extent.' The authors note that:

- dentists, miners, and slaughterhouse and packing plant workers have rates of 'cervical spondylosis' four to eight times higher than other workers;
- welders, sheet-metal workers, and automobile mechanics have ten times the average risk of incurring inflammations in the tendons of their shoulder joints;
- female cleaners are among the groups who face triple the risk of knee arthrosis;
- one electrician or assembler in three has pains in the wrists; and,
- in jobs that require high-strength grips at frequent (less than thirty-second) intervals, the risk for pain in the hands was thirty times greater than for other jobs.[9]

The Swedish National Institute of Occupational Health (NIOH) has pioneered a method of measuring muscle load using electromyography, a videocamera, and a computer. Called PIMEX (Picture Mix Exposure), the method 'is based on making a video of a job while a direct-reading monitoring instrument measures the worker's exposure to a factor in the work environment' by displaying a column whose height corresponds with the muscle load in the activity shown on the videotape. Researchers cut the video to about fifteen minutes and let the worker watch along with a physiotherapist, who can help identify problem areas.[10]

NIOH also organized an international conference to develop strategies to prevent MSIs, with representatives from the United Kingdom, Finland, Denmark, Japan, and Switzerland, as well as Drs Barbara Silverstein and Lawrence J. Fine of the United States. (No Canadian representative is listed.) The working group called for clearer definitions of terms, publicity for the costs-benefits of ergonomics, a good ergonomic environment from elementary school on, MSI reporting in the healthcare system, ergonomic impact studies as well as environmental impact studies for major projects, job variety, and more research. One strategy that participants emphasized was obvious, but often overlooked.

*Ask the Workers*

'Healthy ergonomic conditions in a workplace can not be successfully implemented without input from the employees,' the Swedish working group's report emphasizes. 'The employees' expertise as users of both equipment and tools should be utilised by the employers. It is vital to have active participation from the employees when the improvements are planned. In many situations only the employees understand in detail how a specific work task in performed. In addition, unless a solution is acceptable to the majority of the employees, it will usually be resisted and often not successfully implemented.'[11]

## JAPANESE GUIDELINES

The same philosophy appears in Japanese legislation: 'It is workers who are most affected by changes in the working environment,' reads a section of the Japanese Ministry of Labour Ordinance No. 59, discussing the comfortable working environment. 'In view of this, actions must be taken to reflect the opinions of workers on programs aimed at creating a comfortable working environment by, for example, making active use of the safety and health committee.'[12]

Worker involvement is a hallmark of the Japanese approach to management, of course – and therefore given at least lip service in Western organizations that are hurrying to adopt such policies as just-in-time production and Total Quality Management. Missing in translation, unfortunately, is another hallmark of Japanese efficiency: a determined national campaign to reduce industrial health hazards. The Japan Industrial Safety and Health Association, a department of the Ministry of Labour, promotes a zero-accident campaign that has successfully guided employers all across Japan to safer workplaces.

Japan endured an MSI epidemic in the 1960s and as a result has developed some strong labour legislation regarding working methods and job design, specifically with regard to working with computers. For instance, the 1992 update of the 'comfortable working environment' legislation addressed improved working methods thusly: 'Tasks workers engage in burden their bodies and minds in one way or another. Duties that require unnatural postures and muscular strength are heavy physical and mental burdens for the workers. In view of this, improvements need to be made for these tasks so that the loads are reduced.'[13]

One way to reduce the load is to require employers to direct their workers to take frequent breaks. That's exactly what the Japanese Labour Standards Bureau Notice No. 705, 'Guidelines to Occupational Health in VDT Operation,' does. 'For workers who are engaged in continuous VDT operation,' the work design section directs, 'employers should try to make one continuous operation not more than 1 hour, to provide operation stoppage of 10–15 minutes before the next operation, and to provide one or two short pauses in the process of a continuous operation.' Just to clarify, the Labour Standards Bureau issued a supplemental administrative notice three months later that said, in part, that jobs should be designed to avoid a worker's sitting in front of a VDT all day, and that 'efforts should be made to improve an atmosphere in the workplace so that the [VDT] operators can freely take recesses.'

As well, the Labour Standards Bureau guidelines stipulate that every VDT worker should have a complete physical examination, including ophthamological and MSI tests, before beginning VDT work and then regularly (at least annually) thereafter. Workers should receive at least a half-day training in recognizing and avoiding VDT hazards; their supervisors are required to attend full-day training sessions. The supplemental notice warns that 'it is important to take swift measures when VDT operators complain of any subjective symptoms.'[14]

Although the idea of letting data-entry workers have fifteen minutes off every hour – in effect, barring them from the keyboard or mouse for two hours out of the eight-hour day – might give some U.S. employers heart attacks, it's much less strict than what Japanese unions recommended. Take the Japanese Association of Industrial Health VDT Work Study Commitee's guidelines:

1. No overtime VDT operation should be allowed in principle.
2. Total operation hours should be less than 4 hours.
3. Duration of continuous operation should be less than 50 minutes, with more than 10 minutes' rest in between each continuous operation.
4. Total number of touches on keyboard when mainly inputting figures should be less than 40,000 a day.[15]

---

*Hazard Pay, Anyone?*

'VDT operation is to be regarded as a hazardous job, and periodical physical examinations should be given more than twice a year. Items to be examined include disorders of vision, neck, shoulder and arm, lumbago, radiation injuries, etc.'

— General Council of Trade Unions of Japan (May 1985).

As these examples demonstrate, all around the developed world, departments and ministries of labour standards have identified much the same risk factors and preventive measures for MSIs, although the language and legislation varies somewhat according to the culture of a particular country. 'In Sweden, it's enough for the government to say that workers must have comfortable workstations and variety in their jobs,' reflected Barbara Silverstein, 'where in Japan, the guidelines have to be specific and precise.'

DOWN IN OZ

Then there's Australia where, as in Canada, OHS falls under provincial (not federal) jurisdiction. Australia nonetheless has a National Occupational Health and Safety Commission – a tripartite agency comprising representatives of employers, unions, and governments – which has declared a National Code of Practice for the Prevention of Occupational Overuse Syndrome. As discussed in chapter 2, officially, RSIs don't exist in Australia anymore, or in New Zealand – although residents of both countries participate in Internet mailing lists and discussions about RSIs. MSI is called 'occupational overuse syndrome,' or OOS, there. The National Code of Practice advised employers and junior governments alike:

The expectation of the Commonwealth Government and the National Commission is that national codes of practice will be suitable for adoption by Commonwealth, State and Territory governments. Such action will increase uniformity in the regulation of occupational health and safety throughout Australia and contribute to the enhanced efficiency of an Australian economy ... The application of any National Commission document in any particular State or Territory is the prerogative of that State or Territory.

Legislation in Other Jurisdictions   175

The National Code of Practice checklist for OOS risk factors includes:

- grip maintained for more than ten seconds;
- tasks done for more than one hour at a time;
- tasks done more than once every five minutes;
- similar actions repeated for more than one hour in a work day or shift; and
- similar actions repeated more than several times a minute.

This first checklist is followed by a second checklist for risk assessment, evaluating workstation and job design, with such questions as, 'If fine assembly or writing tasks are performed for most of the shift, is there a lack of support for the forearm?' The question implies the remedy. Here's another example: 'Is there an inadequate number of staff to meet work demands?'

One section of the Code of Practice seems peculiar to Australia: employee duties. The Code specifically requires employees to participate in training, use any mechanical aids or devices provided, take any rest breaks provided, and report to their employer or employee representative any workplace problems involving repetitive actions or awkward postures. Although this section seems to put an onus on workers to prevent MSIs – equally with their employers – perhaps it is intended to protect workers who complain.[16]

BY CONTRAST

Amid such striking similarities in international legislation, and in analysis and identification of risk factors – even in Australia, where the physiological basis of MSIs was once challenged – most amazing of all is the legislative vacuum that exists in the United States and Canada. True, some employers and some industries are already implementing MSI prevention, either out of enlightenment or for the sake of cost containment. True, unions are educating workers about risks and pushing for ergonomic adjustments wherever they can. True, ergonomic consultants are in demand and manufacturers and suppliers are marketing everything from kitchenware to handwear to furniture and manufacturing equipment as 'ergonomic.'

Still, some U.S. and Canadian employers are resisting ergonomics regulation so fiercely, and lobbying so intensely to 'reform' Workers' Compensation, as to raise serious questions about whether they agree at all with the premise that as employers they are responsible for maintaining workplaces that do not pose threats to their workers' safety or health.

# 11
# The Battle over Legislation

'This audience is all civilians watching a national conflict,' mused Frank Mirer, director of the Health and Safety Department for United Auto Workers International. 'And as usual in a conflict, the civilians are the ones who are worst hurt.'

The occasion was the wrap-up panel of the Fourth Conference on Managing Ergonomics in the 1990s (co-sponsored by the Center for Office Technology [COT] and the American Automobile Manufacturers' Association) on the theme, 'A Discussion of the Science and Policy Issues.'

'This conference is a radical departure from previous best practices conferences,' said AAMA health and safety director David Felinski in his welcoming speech. Indeed, four days of plenary sessions featured representatives from business, labour, and government – as well as scientific and medical researchers – in a high-proof distillation of the corrosive battles over Workers' Compensation and ergonomics regulation that have raged in national, state, and provincial capitals for at least ten years.[1]

Although this conference did not address Canadian legislation directly (Canada has no federal OHS authority), provincial WCBs have also contemplated ergonomics regulation and encountered resistance from employers, who use the same arguments as those presented at the conference. With NAFTA, GATT, and other international treaties in force, Canada's labour standards may be harmonized, or called into compliance with other signatories – that is, the United States. Thus, Canadians have a stake in this debate and should be prepared for it to emerge domestically.

Picture sixty highly polished experts presenting state-of-the-art computer-generated slide shows on a Cincinnati hotel stage before some five hundred watchers. Every panel offered at least one speaker from management, one from labour, and one researcher from government or academia. Medical and ergonomics panels featured some of the top researchers in their respective fields, presenting their most recent findings. And all this under the watchful eyes of an audience packed (business registration fell off from previous COT conferences) with about two hundred members from the United Auto Workers' union.

Somehow, most speakers, on panels and from the floor, abided by Felinski's opening request to 'be respectful' of one another. That doesn't mean they didn't disagree, or that there weren't heated conversations in the corridors.

ADMINISTERING SAFETY

The main bone of contention was whether OSHA (the U.S. Occupational Safety and Health Administration) should promulgate any kind of workplace ergonomics standard. In the mid-1980s, a United Food and Commercial Workers' campaign highlighting carpal tunnel syndrome in meat-packing plants led to national boycotts of certain meat packers, and to OSHA's discovery of discrepancies in required record keeping of work-related injuries and illnesses at those plants and elsewhere. In 1990 (then Secretary of Labor) Elizabeth Dole announced OSHA's new ergonomic guidelines for meat packing and said that OSHA would enforce them aggressively. Also, OSHA used its general duty clause to pursue complaints of MSI outbreaks in other industries. Some fines ran into the millions of dollars.

By 1992 figures from the Bureau of Labor Statistics showed such a steep increase in MSIs that OSHA struck a special committee to draft general ergonomics standards both nationally and, by extension, for state-run OSHAs, in the twenty-three states where they exist. The State of Washington Department of Labor and

Industries seconded Dr Barbara Silverstein to head the committee. By the time the ergonomics committee delivered its draft ergonomics standard in 1994, however, the U.S. House of Representatives had been captured by pro-business, anti-regulation Republicans. Remember when the U.S. federal government closed down all so-called non-essential services during the long, arduous 1995 battle between the House and President Clinton over the budget? Among the riders that Republicans tried to attach to the federal budget was one that forbade OSHA to publish or enforce any ergonomics standards.

The rider proposed (repeatedly) by Texas Representative Henry Bonilla stipulated that 'none of the funds made available in the act may be used by the Occupational Safety and Health Administration directly or through section 23(g) of the OSHA to promulgate or issue any proposed or final standard or guideline regarding ergonomic protections, or for recording and reporting occupational injuries and illnesses directly related thereto.'

The appropriations bill that passed Congress on 25 April 1996 read:

'None of the funds made available in the act may be used by the Occupational Safety and Health Administration directly or through section 23(g) of the OSHA to promulgate or issue any proposed or final standard or guideline regarding ergonomic protections.

'Nothing in this section shall be construed to limit OSHA from conducting any peer review risk assessment activity regarding ergonomics, include conducting peer reviews of the scientific basis for establishing any standard or guideline, direct or contracted research, or other activity necessary to fully establish the scientific basis for promulgating any standard or guideline on ergonomic protection.'[2]

That July, Bonilla made another effort to insert his original rider – stipulating that no funds may be used by OSHA for ergonomics. He was again narrowly defeated.[3]

'Many delivery and trucking interests favor the rider,' reported Curt Suplee for the *Washington Post*, 'including United Parcel Service – a frequent target of OSHA action – and the American

Trucking Assocation,' as well as the small-business community and major lobbyists such as the National Association of Manufacturers. At first the anti-regulation lobbyists called themselves the Coalition on Ergonomic Regulation (CER), but later they evolved into the National Coalition on Ergonomics (NCE).

With OSHA effectively stymied (as this is written, Representative Bonilla has introduced another rider forbidding OSHA's standard pending a report from the National Academy of Sciences), the National Institute for Occupational Safety and Health (NIOSH) has become the key government player. NIOSH has identified MSIs as a key area for research and has published several valuable compendiums of existing research, as well as funding new research into workplace links. NIOSH director Dr Linda Rosenstock stated the agency's position at the conference, that MSIs constitute 'a very significant, serious, and largely preventable problem.'

*Locally Speaking*

In mid-1997 California's Occupational Safety and Health Standards Board finally announced its ergonomics standard. It's been a long, bumpy ride, and the controversy is far from over. According to the information package provided with the California Proposed Ergonomics Regulation:

In 1993, the California legislature requested that a state ergonomics regulation be adopted by January 1, 1995. The Occupational Health and Safety Standards Board developed a proposal and conducted hearings in 1994 which revealed much disagreement among affected groups and experts regarding methods to prevent/control cumulative trauma disorders.

The Board decided against adopting the the proposed comprehensive regulation in November, 1994. The January 1, 1995 mandated deadline was missed ...

The California Federation of Labor filed a lawsuit during early 1995 to enforce the state legislature's directive. The court ordered the Board to adopt an ergonomics standard by December 1996 ... On December 12,

## The Battle over Legislation 181

1995, the Board released a new performance oriented ergonomics regulation proposal.[4]

In April 1996 an editorial in the *Los Angeles Times* endorsed ergonomics regulations on the grounds that 'hundreds of thousands of workers are now afflicted with severe and often crippling pain in their hands, wrists, shoulders and backs from performing the same task over and over. Some specialists dispute the Labor Department's figures, but all agree that the problem is growing ... As those companies that have moved aggressively to reduce RSI can attest, inaction is penny-wise and pound-foolish. The Labor Department should act quickly and business should support it.'[5]

When the ergonomics standard finally cleared the language clarity requirements of the California Office of Administrative Law, at the beginning of June 1997, it ran into new obstacles: lawsuits from the American Trucking Association (challenging the constitutionality) and the AFL–CIO, which wanted a stricter standard.

BURDEN OF PROOF

A 1994 brief item in the *Wall Street Journal* said that the National Association of Manufacturers opposed OSHA's proposed standard as 'programs based on unproven medical and scientific theories.'[6] This is an interesting tactical position, which apparently puts the burden of proof on the regulation advocates.

However, as the Managing Ergonomics conference unfolded, expert after expert described solid evidence of the links between repetitive work, awkward body position, forceful action, and MSIs. Dr Linda Rosenstock reviewed NIOSH data to show that 'research has demonstrated the link between job factors and [CTD] incidence.' Others included:

- Dr J. Steven Moore, who described the links between certain work positions and certain upper-extremity MSIs;[7]
- Dr Alfred Franzblau, who presented results of a survey that

videotaped 352 workers and scored their jobs on ergonomic factors, and found a 'highly significant linear correlation between injury and repetition';[8]
- William Marras, PhD, who reported on an ongoing project that has collected data on thirty-five million person-hours of work to date, and which resulted in a five-variable chart that can reliably predict which jobs will cause back injuries;[9]
- Don Chaffin, PhD, who described another reliable formula for predicting hurt backs, using biomechanical models, called the 'maximum moment';[10]
- Dr Kurt Hegman, who presented the 'strain index' for jobs involving upper limbs, a product of six multipliers, in which a score smaller than five is safe, and a score greater than five is hazardous to upper limbs.[11]

DR LOUIS OBJECTS

Dr Dean Louis presented the position of the American Society for Surgery of the Hand (of which he was then president), namely, that there is no proven link between occupation and upper-extremity disorders. He blamed 'spurious epidemiological research' and lamented that doctor and patient have been 'polarized' because 'patients have been told that they have certain conditions.' But he had to acknowledge – in response to a question from the floor – that he sometimes tells a patient to stop work in order to heal.

ONE HAND CLAPPING

Although the American Society for Surgery of the Hand (ASSH) has taken a position that RSIs do not exist, a sister organization, the American Association for Hand Surgery (AAHS) held its *third* symposium on CTDs in August 1996. The January 1997 issue of the *Journal of Hand Surgery (American Volume)* offers a thorough review of the professional debate, partly because the editor invited two opposing articles, and partly because of the lively letter section.[12]

In the first article, Dr Susan Mackinnon and physiotherapist Christine Novak argued that 'the terms *overuse syndrome* and *RSI* have been used in orthopedic literature to describe athletic injuries, and yet confusion arises when these terms are applied to work-related injuries ... for the worker, we are asked to prove that these disorders exist, to establish blame and determine causality ...'[13]

In the next, Dr Nortin Hadler argued that, 'we are at the point today to assert that any epidemiological investigation purporting to address the incidence or prevalence of or causal associations with median neuropathy at the wrist that does not rely on electrodiagnostics to define the health effect is junk science!'[14]

---

*Laying Blame*

'Dr Kasdan and Dr Ireland [of Australia] have variously blamed RSI on whining patients, greedy lawyers, job-seeking ergonomic experts (and their graduate students), self-serving media, industrial rehabilitation specialists, trade unions, physical therapists, and even ergonomic furniture designers ...[15]'

— Letter to the editor, *Journal of Hand Surgery*.

---

ANOTHER CAUSE?

Actually, National Economic Research Associates' president Mark Berkman suggested another cause for the increase in reported MSIs. Presenting a study that NERA researched on behalf of the National Coalition on Ergonomics, Berkman offered a graph that purportedly showed a rise in WCB benefits paralleling the increase in reported MSIs. He suggested that another cause might be unions and OHS agencies that present free workshops on CTS, which creates 'lots of incentives for people to be aware of ailments they didn't know they had.'

Unfortunately, Berkman's WCB chart isn't reproduced in the précis of his report which appears on the American Trucking

184 Wounded Workers

Association website. But as we have seen in chapter 5, the trend in WCB benefits has been downward in the 1980s and 1990s. At best, his data must be regarded as selective.

*Cost-Benefit Analysis*

Berkman also looked specifically at ergonomics regulations costs and benefits to members of the American Trucking Association (ATA). He estimated $6.5 billion in costs, and called this equivalent to the costs of not meeting the Code – e.g., WCB benefits and fines, medical costs, and litigation – which on the face of it indicates a serious injury rate. The NERA/ATA study estimates that new rules would cost $2.4 billion in increased personnel costs.[16]

If the ATA seems reluctant to see truckers work shorter shifts, other groups are eager to see that job change. 'Fatigue and sleep deprivation are the norm for drivers behind the wheel of large trucks,' warned Citizens for Reliable and Safe Highways (CRASH) in a May 1997 press release. CRASH blamed fatigue (along with the increasing size of rigs) for the almost 5,000 truck-related highway fatalities in 1995: 'A 1989 study of long-haul truckers driving from Washington to Minnesota concluded that at least 58 percent of those trucks violated hours-of-service rules,' CRASH claimed.[17]

CRASH has a Canadian affiliate – Canadians for Responsible and Safe Highways – which reported in February 1997 that 'there is a one in three chance that the big truck you meet on the highway has a mechanical defect serious enough to be placed out of service. There is also a chance that the driver is fatigued and under economic pressure to drive as much as 13 hours a day just to make a living.'[18]

WHAT ARE THE ODDS?

Conservatives are opposed to much more than just ergonomics regulations, of course. They're mounting wholesale attacks on regulations of every kind, from occupational safety to environmental protection to airplane inspections. To conservatives, the term 'risk assessment' has become a pejorative, and they have held the entire discipline up to ridicule.

Nay-sayers like to mock the odds presented by risk assessment. 'There is not a single thing you can do in an ordinary day – sleeping included – that isn't risky enough to be the *last* thing you ever do,' deadpanned Jeffrey Kluger in 'Risky Business,' an article for the May 1996 issue of *Discover Magazine*. Mr Kluger is the author of the book on which the hit movie, *Apollo 13*, was based. In this tongue-in-cheek piece, he rhymed off these statistics: a 1 in 7,000 risk of shaving injury, a 1 in 5,800 risk of dying in a motor vehicle accident in any given year, a 1 in 37,000 risk of an office worker's dying in a job-related accident. He barely slowed down to mention that workers in the poultry-slaughtering industry face a forty-three per cent annual risk of serious injury, and followed this information with a joke that for the poultry, the death rate 'clock[s] in at a cool 100 per cent.'[19]

Countering this viewpoint comes the thoughtful voice of Adam M. Finkel, director of Health Standards Programs at OSHA, and a pioneer in the field of risk assessment. Finkel actually presented his defence of risk assessment in a talk at the University of Calgary School of Management Studies, and mentioned then that he was developing it as a piece for *Discover*, where he wrote: 'The version of QRA [Qualitative Risk Assessment] that many in Congress, academia, and the media have embraced is a repudiation of much of what has gone before in this field. It is reform premised on a myth – namely, the myth that current assessment methods routinely exaggerate risk, at a huge cost to society ...'

Of course, the subtext here is finances. Up until now, workstations have been designed to be functional, to be easy to manufacture and distribute, and to fit into capital spending plans. Fitting the equipment to the worker has come as an afterthought. When legislators and CEOs balk at risk assessment, they are really asking whether they could save money by ignoring the problem, at least temporarily. Accepting the survey figures of CTD*News* for a moment (and we don't know whether most CEOs do accept them), if only one out of eight workers gets injured, then a CEO might hope that all his or her employees are among the other seven.

'Suppose you are told that the average amount of time you need to get to the airport from your house is 20 minutes,' suggested Finkel, 'but that the drive could take as little as 5 minutes or as long as 80 minutes. If you would rather be 4 minutes late for your plane than 5 minutes early, then the average estimate is the one for you. But for those of us who regard missing the plane (or allowing pollution to cause some unnecessary deaths) as more dire than having to wait a few minutes (or wasting some money on pollution controls that turn out to be overly stringent), a more prudent estimate is called for.'[20]

Of course, there's another dynamic here as well, one that has to do with political beliefs and economics systems. As Finkel wrote, 'Above all, we should not pretend we are promoting 'good science' when we are really pushing a political ideology – one that says less government regulation, at least where health and the environment are concerned, is always better than more.'[21]

SELF-SUPPORTING

Presentations at the Managing Ergonomics conference reinforced results reported by CTDNews and elsewhere: as a general rule, ergonomics programs *at least* pay for themselves and often result in substantial savings or extra profit. At Ford, at Red Wing Shoes, at Aetna Insurance, at an unnamed New York garment factory – ergonomics changes have paid for themselves quickly.

There are caveats: finding the right approach takes time; involving all the stakeholders requires sincere outreach; the reported MSI incidence usually shoots up at first as the severity of reported problems drops; and the program needs to be monitored and adjusted regularly. With all those givens, however, most employers do achieve satisfactory results, measured both in reduced MSIs and in increased productivity. Meat-packing plants seem particularly difficult to turn around – but then most meat-packing plants keep speeding up the line.

ORDINARY ACHES OF AGING?

Another argument against ergonomics regulations is that at least some of the aches and pains occur in people who are not involved in job-related activities that have been identified as risk factors. Attorney Terrence H. Murphy articulated some of the questions raised on behalf of one of his clients, a major healthcare concern facing OSHA action about hurt-back claims: 'All of us lift sometimes,' he said, 'and all of us suffer back pain sometimes.' His list of problems with lifting regulations included: back pain is not an identifiable injury; often there's no precipitating cause; everybody gets it some time; the pain level is strongly influenced by psychosocial factors; and low-back pain is hard to research.[22] This last point, of course, brought an energetic response from the back researchers in the room.

Attorney Joseph D'Avanzo, who has successfully defended both IBM and UPS against MSI claims, took a somewhat safer route when he asked: 'How come nobody integrates into our model what people do when they're not at work?' He said he didn't believe his clients should be liable for disorders that people would have developed anyway.[23]

Barbara Silverstein responded succinctly, 'It's like smoking. Some people can smoke for twenty years and not get sick. That doesn't mean tobacco's safe.'

Tashlyn Chase, a Canadian Auto Workers member and Ford Canada's National Ergonomics Coordinator, pointed out that there's a difference between work and leisure. 'When something hurts at home, I can stop doing it,' she said. 'At work, I don't have that option. When workers go home, we go home tired. We hurt so much we can't pick up our kids. The onus has been placed on us for far too long.'

---

*Asking Too Much?*

Long-time ergonomist Dr Suzanne Rogers suggested a simple and sensible guideline for workplace design: 'I think we ought to design our jobs so that people can lead normal lives outside of work,' she said.

---

LEGAL LANGUAGE

'Presence of injury does not prove presence of a hazard,' insisted attorney and NCE spokesperson David Savardi. He argued that OSHA's draft ergonomics code had to be discarded because it failed to meet regulatory requirements. He presented a discussion with slides which he apparently uses when teaching at a law school, and which purports to outline the legal requirements for OSHA regulations. He argued that in order to prove significant risk exists, the agency must produce background-level data (occurrence in the general population), and must show that the proposed measures mitigate risk to a significant degree, and that the means of abatement are feasible at a feasible cost. He argued – with great assurance – that scientific issues must be resolved before regulations can be adopted.

Later on the same panel, session arranger (and COT Executive Director) P.J. Edington introduced Randy Rabinowitz, Director of the Project on Federal Regulation at the University College of Law. In her clear, carrying voice, Rabinowitz reviewed a list of recent court decisions on OSHA regulations and enforcement practices, almost all of which contradicted Savardi.

'No proof of a causal relationship is required,' she said. 'There's no requirement for dose/response data. Courts have consistently rejected arguments about non-occupational factors. The Court has ruled OSHA is supposed to regulate in the face of scientific uncertainty and not await the Godot of scientific certainty.' In particular, she said, the courts have specifically prohibited cost-benefit analysis as a criterion for regulations. Point by point, she followed Savardi's purported requirements – and on practically every point showed that Savardi was wrong.

### THERE OUGHTA BE A LAW

'You can identify high-risk industries and high-risk jobs,' said Dr Lawrence J. Fine of NIOSH, confidently. Law professor Sidney Shapiro seconded that: 'The science argument is a policy argument,' he said. 'It's not relevant to OSHA's mandate.'

'All the methods discussed here [at the conference] are readily available,' said Barbara Silverstein. 'All of the checklists boil down to ask, look, and measure. Beyond that, there's a continuum of precision and difficulty.' The purpose of the OSHA ergonomics standard was to identify high-risk jobs so that employers and workers could protect themselves. 'This is a policy decision,' she said. 'There are 6.2 million workplaces and not enough ergonomists. Our goal was triage, a tool that was sensitive, rapid, and easy to use.'

Whether OSHA will be able to promulgate such a code in the current political climate is a moot question. Peg Semanario of the AFL–CIO observed privately that OSHA will continue to investigate employers and enforce the general duty clause anyway. NIOSH continues to generate research and guidelines for employers to adopt voluntarily. State OHS agencies, such as those in California and North Carolina, have demonstrated keen interest in ergonomics violations. Even the most recalcitrant U.S. employers may have so many examples of successful ergonomics programs around them that eventually they'll begin to request some kinds of standards. But that's far from a safe bet.

Rather, what seems to be happening (in Canada as well as the United States) is that employers are moving to new procedures for retaining workers. Instead of actually hiring people, they're taking on part-time, contract, or contingency workers – who have no claim on full-time benefits, and for whom they can deny any health and safety responsibility.

# 12
# By the Fingernails

HANGING ON TO JOBS

Of all the catch-22s involved with MSIs, perhaps the bitterest is that the injury makes the worker view the job that injured her as a lifeline, as her key connection with the workforce – but makes the employer view the same job as a liability, to be downsized or divested or outsourced at the earliest opportunity. That means that MSIs (and other OHS issues) could be the swift kick that politicizes people, not only the injured worker but also her family and community.

Rampant downsizing during the 1980s and 1990s has made individual employees appear – and feel – expendable, even when able-bodied and in good health. 'Seemingly overnight, the old idea of a job has begun to seem like a social artifact. The new idea of rapid turnover of jobs, constant retraining, insecurity and flat wages has transformed the way millions of Canadians live, organize and enjoy their lives,' warned an article from Southam News.[1]

'More than 43 million jobs have been erased in the United States since 1979,' the *New York Times* reported in March 1996.[2] And although labour statistics show a net increase of twenty-seven million jobs in that same time period, the new jobs are at much lower wages. 'The result,' according to the *Times*, 'is the most acute job insecurity since the Depression.'

The *Wall Street Journal* reviewed six growth jobs in the 1990s and titled the resulting article, 'Nine to Nowhere.' Writer Tony

Horwitz concluded that 'while American industry reaps the benefits of a new, high-technology era, it has consigned a large class of workers to a Dickensian time warp, laboring not just for meager wages but also under dehumanized and often dangerous conditions.'³

Nor is education the job ticket that it once was: 'Workers with at least some college education make up the majority of people whose jobs were eliminated,' noted the *New York Times* 'and better-paid workers – those earning at least $50,000 – account for twice the share of the lost jobs than they did in the 1980s.'

'Companies everywhere are downsizing to maintain a core of permanent employees supplemented by an outer ring of disposable workers on temporary or short-term contracts,' is how writer Victor Keegan put it in the *Guardian*.⁴ Making the same point, the *Globe and Mail* analysed a Statistics Canada report and found that 'last year [1995], only 54 percent of workers put in a standard workweek, down from 65 in 1976 ... The shorter hours available to relatively unskilled people in some industries like hospitality suggests that such workers "may be treated as roughly interchangeable."'⁵

Of course, employers treat disposable or interchangeable pieces differently than they treat valued employees. Not only has job security all but disappeared, but in many cases, so has the actual act of hiring an employee, with all the consequent legal obligations. Employers offer piecework instead of minimum wage. They avoid paying benefits by hiring two or three part-timers instead of one full-time worker – or they hire full-timers on contract, as so-called independent contractors, who are responsible for their own health insurance and disability coverage. If a worker gets sick or hurt or can't do the job, the company is in a position where it can legally let the worker go, without paying severance or vacation or pension or – you guessed it – disability benefits.

## REVISING THE CONTRACT

'My great concern,' said occupational physician Brendan Adams, 'is that we seem to be entering the era of the throwaway worker.'

The trend to contract work, temporary work, work-from-home, part-time work and other non-normative forms of employment threatens to have very serious consequences for people who develop MSIs or other work-related diseases or injuries, because employers will have little or no legal responsibility either to reduce MSI hazards or to compensate or rehabilitate injured workers.

The Canadian Labor Congress Health and Safety Committee sees a bleak future for injured workers: 'Workers' compensation, in all jurisdictions, has been eroded to the point that it cannot adequately address the effects of work today. More and more injured or sick workers are finding themselves forced on welfare or other disability programs. Now, the entire social program system is being reformed, putting these people and their already shaky physical and economic viability at further risk.'[6]

As the workforce divides, willy-nilly, between the overworked and the unemployed, occupational health hazards such as MSIs and stress – most particularly, MSIs – may be the key factor in determining whether employers can shape the workplace entirely to suit their own profit margins, or whether they will be forced to take human limitations into account.

WHITHER THE WORKFORCE?

Wise managers have always realized, of course, that human resources are any company's most important assets. And, as Henry Ford maintained, workers need sufficient wages to buy the company's products as well. Yet the whole thrust of capitalism has been to replace human assets with capital assets – to invest in machines that allow companies to keep lowering the cost of labour by breaking jobs into simpler and simpler components that can be done for lower and lower wages.

---

*Work for Sale*

'Labor power has become a commodity,' as Harry Braverman reaffirmed in 1974. 'Its uses are no longer organized

> according to the needs and desires of those who sell it, but rather according to the needs of its purchasers ... Every step in the labor process is divorced, so far as possible, from special knowledge and training and reduced to simple labor. Meanwhile, the relatively few persons for whom special knowledge and training are reserved are free as far as possible from the obligations of simple labor. In this way, a structure is given to all labor processes that at its extremes polarizes those whose time is infinitely valuable and those whose time is worth almost nothing.'[7]

An assembly line is one example of the job's being removed from special knowledge and reduced to simple labour. Unlike the butcher, who must know how the carcass is constructed in order to take it apart neatly, the typical meat-packing worker needs only to know how to remove one section of useable meat (although that worker may be promoted to another part of the carcass in time). Professor Michael Broadway says meat packing changed fairly recently, according to a *Calgary Herald* report: 'In the 1960s, a company called Iowa Beef Packers emerged [that] replaced the once relatively skilled, well-paid job of a meat cutter with a low-paid laborer rate more typical of an assembly line worker.'[8]

A typical electronics assembly worker may understand how to connect only one circuit, or two, or three. At the auto plant, the typical worker spends all day assembling, say, wheel wells, as they roll by – or putting in the seats, or the windshields. In every case, the worker's skill is minimized, and therefore the worker's pay package is lowered.

CLERICAL ASSEMBLY

The advent of computers, however, has added some wrinkles to this trend towards deskilling work. Where once managers dictated letters for secretaries to type (whose time was cheaper), now managers can peck out rough drafts on their own computers – or choose boilerplate paragraphs – and let their secretaries clean up the drafts on diskette. With documents on diskette, even the lengthiest contracts can be revised and printed as quickly as the parties involved can renegotiate the terms. One secretary can do the work formerly done by several.

Supermarket cashiers have also seen computers expand their jobs, so one person now does jobs formerly performed by three or four workers. The cashier not only enters the prices and totals them, she or he also bags the groceries, updates inventory records, and weighs produce and bulk goods. To someone who's never really looked at the job – much less done it – all a cashier seems to do is to slide packages across the scanner window (or run a pen over the UPC). The notion that she (or he) should need frequent breaks, or be paid more than ten dollars an hour, seems absurd to some people.

Thus, an economics professor such as Paul R. Krugman of Stanford University can write a serious article in the *New York Times Magazine* with the straight-faced assertion that 'even the fanciest information technology may have only a marginal impact on the material world in which we all still live ... bar codes and laser scanners are nifty, but there's still time to read about celebrities and space aliens in the checkout line.'[9] Obviously, an economics professor lives in a different kind of material world from a cashier's – a world freed as much as possible from 'simple labor.'

THE END OF JOBS?

Other experts take an opposite view. 'We are, indeed, entering into a new period in history – one in which machines increasingly replace human beings in the process of making and moving goods and providing services.' So asserted social commentator Jeremy Rifkin in his 1995 book, *The End of Work*. He traced in some detail the growth of automation in farming and agriculture, in manufacturing everything from autos to textiles, in office work, and in service industries. He found that the net effect of automation is higher profits for corporations and shareholders, but fewer jobs and benefits for workers. And as jobs disappear, capital becomes more and more concentrated in the hands of a very few people at the top.

This trend, Rifkin argued, results in a polarization that is socially and economically destablizing for both the developed and underdeveloped world. 'The net worth of the [U.S.] nation's 834,000 richest families now totals over $5.62 trillion,' he pointed out. 'In contrast, the net worth of the bottom 90 percent of American families is only $4.8 trillion ... Less than half of 1 percent of the American population now ... owns 37.4 percent of all corporate stocks and bonds and 56.2 percent of all U.S. private business assets.'[10] That is, would-be workers lack not only access to jobs and earned income, but to resources of all kinds.

Bruce O'Hara adds that not only do machines produce most of the goods we need, they actually produce more than we can consume. 'Productivity expands more readily than consumption,' he wrote in *Working Harder Isn't Working*. 'This is the wall against which the modern economy breaks, time and time again.'

Personal and environmental limits to consumption are key factors in creating what O'Hara called a labour-surplus society:

Unworkable economic strategies have created a horrible paradox: the more efficient our technology becomes, the more inefficiency it injects into the econonmy as a whole. Each time I, as an individual, work longer and harder, it makes the national economy less efficient and less productive.

Inefficiency is not an intrinsic feature of the market mechanism. When kept in its proper operating range, a market economy is very efficient. Inefficiency builds up whenever the economy is designed in such a way that production chronically outstrips consumption. We can state this limitation on a market economy's operating range in the form of a design constraint: To function efficiently, a market economy must be structured so that an approximate balance between production and consumption is always maintained.[11]

## MEN WITHOUT JOBS

Men without jobs tend to fall into antisocial behaviour. This conclusion was presented by two very different sources: William Justin Wilson in his book about African-American neighbourhoods (*When Work Disappears*) and the *Economist*, which reviewed the effects of automation on the European workforce. Wilson pointed out that paid work gives structure and purpose to daily life; without it, men are at loose ends. One in three African-American men has been or is in conflict with the law.

The *Economist* fretted that 'men do not necessarily adopt "social behaviour" (obeying the law, looking after women and children) if left to themselves; rather they seem to learn it through some combination of work and marriage ...' Yet, changes in the workforce mean that more and more men do not have jobs – often for long periods of time. 'The labour market is increasingly friendly to women,' according to the article, '(though men still make more money and are more likely to have jobs); but there are growing numbers of men outside the labour market in a way that women have been accustomed to but men are not.' Among women's advantages in the labour market: 'their willingness to accept lower-paid jobs.' And when men cannot provide, they retreat into 'fundamentalist masculinity – the world of gangs.'[12]

Wilson recommended a new WPA, a national public works project that would rebuild the U.S. infrastructure including the health and education systems, perhaps create an artistic renaissance, and above all provide meaningful paid activity to people who are now idle and without hope or self-esteem.

WOMEN'S WORK IS ...

What's missing from all these accounts is, of course, any recognition that the effect of what Rifkin calls the Third Industrial Revolution is different for women – no less severe, but different. Jobs traditionally held by pink-collar workers have been hit hard by technology, deskilled, devalued, doubled up or eliminated. 'The number of women with multiple jobs has increased since 1976 from about 50,000 to more than 300,000,' Southam News observed.[13] And that's in the *paid* workforce – not even counting the double shift of domestic work that waits at home. Even *The*

*Economist* noted that men without jobs are still reluctant (not to say recalcitrant) to take up domestic chores.

Worse, most large corporations are flattening their hierarchies, eliminating the layers of middle managers who are no longer needed to ensure that top management's decisions are carried out. Since the 1970s, women have been pushing themselves to get into middle management, only to bang their heads against the 'glass ceiling.' Now the glass ceiling has been lowered, and most women's prospects have been lowered too.

Ironically, at the same time that large corporations have been eliminating jobs (by the thousands and tens of thousands), business lobbyists have won huge cutbacks in government payrolls, with the claim that an unfettered private sector could create more jobs more efficiently than the public sector. Spending cuts have hit all parts of public services – education, healthcare, Workers' Compensation, seniors' benefits – but cuts to the social safety net have been especially deep.

'The attack on waste and duplication in government has turned into an all-out assault on public programs at every level,' Angus Reid wrote in his 1996 book, *Shakedown*. 'A combination of privatization, budget cutbacks and steep user fees is eroding the foundations of Canada's public infrastructure. There is no longer a cost-benefit analysis when it comes to providing public service – only a cost analysis ... What are we to make of the new knowledge-based economy when a Vancouver hospital anounces it is cutting nursing positions so it can hire less-trained and lower-paid nursing assistants to do their work?'[14]

Cutbacks also increase the amount of unpaid work consigned to the household, such as home schooling or extra tutoring for students, extra caregiving for sick family members, unpaid childcare for relatives who have to find paid work, and juggling family finances to compensate for cutbacks in drug plans and complementary medical care.

Then there's 'workfare,' which presents an extreme catch-22. Workers who develop severe work-related MSIs now face limitations on WCB benefits in many provinces and states. When WCB payments run out they have to apply for social assistance. And in order to get social assistance, in some if not most jurisdictions, they have to perform some kind of work. In Alberta, as in the United States, tight restrictions on eligibility for social assistance and disability benefits have already led to visibly increased numbers of women and children living on the street. Only time will tell how such families will survive. Social workers will be watching crime statistics apprehensively.

Maybe MSIs are just the growing pains of changeover to the Information Age. New industries do bring new job hazards – such as telegrapher's wrist – and in the past most of them have been ironed out eventually. Maybe in twenty years or so, the MSI problem will just go away, either as a side effect of other changes or because ergonomics offers cost-wise practices. Even if that were the case, that would still leave sizeable numbers of workers at immediate risk and patients in need of immediate help.

Unfortunately, such a natural and inevitable transition does not seem imminent. With governments backing away from OHS enforcement, unemployment high and job insecurity rampant, we may be in for lessons in how the so-called free market treats workers. Of course, most of these lessons have already been described by such authors as Charles Dickens and Upton Sinclair.

SAVE OUR HANDS!

In his review of occupational health hazards, Allard Dembe concluded that the United States needs:

- a universal healthcare plan;
- a universal disability insurance plan;
- the Occupational Safety and Health Administration and other regulatory agencies to enforce strict OHS regulations; and
- epidemiologists to monitor the workforce to identify hazards.[15]

Although comprehensive, and perhaps idealistic in the context of U.S. politics, this plan still lacks any apparatus to deal with the unequal division of (unpaid as well as paid) work, the increasingly mechanized and deskilled structure of jobs, the trend away from formal workplaces or employment agreements, or the question of what happens to workers who are already injured.

RESPECT OUR SKILLS!

The Swedish National Institute of Occupational Health (NIOH) has addressed the second point with a recommendation that jobs that have been deskilled should be re-skilled. 'As a general principle, an employee who works with data input should also take care of data output,' according to NIOH's Professor Gunnar Aronsson. 'This reduces the feeling of isolation. There is also a psychological advantage in letting the employee personally experience the consequences of bad data quality.'[16]

On this side of the Atlantic, re-skilling occurs mainly by accident. Greg Hart remembered what happened when he set up a rehabilitation program for injured workers. 'We had them building phones by hand, from scratch,' he said. 'And you know what? They built them faster than the assembly line!'[17]

The urge to do the whole job, a complete job, to take a project from start to finish and to be able to take some pride in completion – that may be part of the reason that so many people (especially women) are choosing to start their own businesses rather than hunt for corporate positions.

Another aspect of the Swedish experience is that using ergonomics to adapt everyday objects 'can lead to better products for everyone,' according to a paper by Maria Benktzon in *Applied Ergonomics*. Benktzon looked at prototype ergonomic adaptations of eating implements, walking sticks, and coffee pots and found that 'addressing particular aspects of design for people with specific difficulties and problems ... has led to designs which are acceptable to a broader range of users.'[18] Think about it: if workplaces have to be redesigned to accommodate new equipment, that's the perfect time to build in changes to accommodate persons with disabilities, including MSIs.

H IS FOR HOME, J IS FOR JOBS

Of course, many employers are looking to do away with centralized workplaces as much as they can, using alternatives such as telecommuting, working from home, or piecework. Employees carry out most of their work-related activities from their own homes and come in to the centralized facilities only occasionally, mainly for meetings. While this trend is obviously more environmentally friendly than having hordes of people descend on downtown at the same time in the morning and then leave for home around the same time at night, there are other, troubling aspects as well.

For women especially, the social aspect of employment is very important. Workers may be more productive because of the ease with which they can casually swap ideas in a shared workplace. Also, although fewer employers say it out loud anymore, work in the home has in the past been promoted as a convenient alternative to workplace childcare – convenient for the employer, that is.

Contingency workers, whether for factories or supermarkets, find that they always have to be 'on call' in order to be available for the few hours a week their employers need them, says sociologist Louise Vandelac. 'A large part of their problem ... was irregular working hours, a problem especially acute among telephone operators, who had a different schedule every week, sometimes every day, making it impossible to plan family activities or have any kind of routine at home.'[19]

Where MSIs are concerned, the home workplace could mean that individual workers assume (perhaps unwittingly) both the expense and complete responsibility for creating an ergonomically safe workstation – regardless of whether they have the resources to do so. As an example, pieceworkers in the textile trade often labour in poor lighting, on elderly machinery, for stretches as long as they can manage in between or after their family duties.

TEMP-ERING THE STRAIN

Manpower Temporary Help agency is now the largest single employer in the United States. Temporary workers swing from workplace to workplace, applying themselves intensely to push through large projects, working from whatever (usually non-adjustable) workstations happen to be available at the time. Who bears responsibility if the temp worker should develop an MSI – the agency that hires out, the company that hires in, or the worker? Such workers need to be represented somehow, and accorded the protections that are now regulated standards in permanent positions and in regular workplaces.

SAVE *OUR* HANDS, TOO

As employers clamour for deregulation of employment practices, that opens up the whole question of what sorts of guidelines are appropriate for the evolving workforce. Canada and the United States need to examine a whole range of issues, such as:

- tripartite labour market planning, by national or local councils where representatives of labour, employers, and government all meet to discuss current and future needs, as happens in most European countries;
- revisions to NAFTA, APEC, MAI, and all other international trade agreements to include minimum OHS codes, along the lines of the European Union model, rather than encouraging the so-called free enterprise zones (where labour and environmental regulations do not apply) now used to attract foreign-owned manufacturing plants;
- stipulating that all trade treaties shall abide by all ILO conventions;
- international research teams, working to identify and alleviate workplace hazards and otherwise to develop safer OHS standards;
- international medical research teams to develop effective, non-invasive treatment modalities for patients with severe MSIs.

Although many people who live in the United States envy the Canadian healthcare system, Canadians with disabilities might envy the U.S. *Americans with Disabilities Act* – especially now, when Canadian programs are being discontinued or decentralized, benefits are being cut back, and eligibility is being tightened at every level of government. A Canadian disabilities act would provide a much-needed ramp to self-sufficiency for Canadians with disabilities, including MSI patients. Also, on the OHS front, while a federal investigative and enforcement agency seems ever more unlikely, Canada might borrow from Australia the strategy of developing a national code of practice – in effect, federal standards – for ergonomics, and pressuring provincial governments to adopt them in their own jurisdictions.

SIX HOURS A DAY

But the core issue, the question at the heart of the MSI problem and at the centre of the whole shift in the workforce, has to do with *hours.* How many hours a day can people use their hands without seriously injuring themselves? How many hours do people have to put into paid employment in order to maintain themselves and their families in security and comfort and a modicum of freedom?

Wages have actually been dropping during the 1990s – not their buying power, but wages themselves – even as jobs become less secure, making people more willing to accept overtime. Mandatory overtime not only injures people, it wreaks havoc with family life too. In the 1996 Canadian Auto Workers' strike against General Motors, mandatory overtime was the issue on which workers stood their ground and would not budge. Yet, one reason for the high unemployment rate is that employers prefer overtime to hiring more workers.

A 1995 Angus Reid poll found that 'about two-thirds of all Canadian workers reported that they were working longer hours compared with a few years ago; only about a quarter said they were getting paid for this extra effort,' Reid reported in *Shakedown.* In 1996, 'about thirty percent of Canadians told us they felt that they, or someone in their family, would probably be laid off or become unemployed in the next twelve months.'[20]

Bruce O'Hara reviewed a wide range of solutions to the job insecurity conundrum and concluded, 'Reducing the length of the standard workweek has been the simplest, most elegant solution to problems of overcapacity. Not only does a shorter workweek restore economic balance in times of surplus, but it is also a benefit to the workforce in and of itself. Extra time for family life and leisure is a boon for working people.'[21]

What workplace automation could buy us – all of us – is more time to pursue interests other than paid work. Unions are already calling for shorter workweeks, and especially shorter workdays. Union activism was key to the adoption of the ten-hour day in the early nineteenth century, and the eight-hour day circa 1917. Then, as now, machines made greater productivity possible in fewer hours, thus eliminating jobs. Now, however, too few new jobs are appearing in new sectors to absorb the displaced workers.

According to Jeremy Rifkin, the U.S. Congress came very close to legislating a thirty-hour workweek – as a job creation measure – as long ago as the Great Depression. 'Much to the surprise of the country, the Senate passed [Senator Hugo L.] Black's bill on April 6, 1933 ... mandating a thirty-hour week for all businesses engaged in interstate and foreign commerce,' he wrote.[22] Even though the bill was subsequently killed in the House, cutting the workweek was already seen as a patriotic gesture.

> *Thirty Hours a Week – in 1932*
>
> 'Major employers, including Kellogg's of Battle Creek, Sears, Roebuck, Standard Oil of New Jersey and Hudson Motors voluntarily cut their workweeks to keep people employed,' wrote Jeremy Rifkin. 'A survey of 1,718 business executives conducted by the Industrial Conference Board found that by 1932 more than half of American industry had reduced the number of hours worked, in order to save jobs and promote consumer spending.'[23]

Unfortunately, too few large corporations these days feel any sense of social obligation or even have roots enough in any one nation for an appeal to patriotism to be an effective inducement for reducing the workweek. And considering the way that corporate profits have soared, offering tax breaks for companies that shorten their workweeks seems problematic on several levels. For starters, shorter hours are actually good economic as well as sound ergonomic practice. There's some evidence that employees who can choose their own hours are more productive than those who have to fill their allotted time at the workplace.

Shorter workweeks all around would offer obvious appeal for women. As caregiving work gets pushed out of the paid workforce back into the home, women find the burden of employment and family duties heavier and heavier. With shorter workweeks in place, not only would women have more time for family life, but so presumably would their partners – more time to dandle babies, help with children's homework, or put up homemade jam, as well as run errands and maintain household machinery.

Are Canadians (and Americans) ready to pass new labour laws that require major employers to shorten the workday, and hire more workers instead of demanding overtime from those already in place? Oddly enough, public opinion seems to be shifting away from the corporate philosophy.

INQUIRING MINDS FIND TRENDS

In Canada, political scientist Neil Nevitte has found that 'our fundamental orientations towards authority have changed in the polity, in the workplace, and in the family.' Nevitte's study of Canadian values over a fifteen-year period – and comparison with U.S. and European values – sketched a portrait of people becoming what he called 'post-materialist.' Along with a growing environmental consciousness and more egalitarian family life, post-materialism means 'being more interested in having a job that fills your abilities, meets your interests, satisfies your creative instinct rather than a job that provides you with security ...'

In short, Canadians in the 1990s are more willing than they have been since the Second World War to confront hierarchical assumptions. Nevitte found that, with such behaviours as face-to-face political discussions, boycotts, and other kinds of low-level political protest, 'Canadians are far more likely to [participate] than their American counterparts.'[24]

Even more surprising, a public opinion poll taken in August 1996 found that Americans are 'fuming about recent corporate behavior.'

Specifically, voters are angry about the following kinds of behavior:

- Massive layoffs at a time when profits and CEO salaries have been going through the roof;
- Firing full-time workers and filling their jobs with temps who get neither medical benefits nor pensions;
- Firing older workers and replacing them with younger workers satisfied with lower wages.

In sum, people are angry that corporations are showing disrespect for, and disloyalty to, the American workforce.[25]

Even though the 800 survey respondents said they were still angry at governments, there was a stunning turnaround: almost seven out of ten 'say they generally favor government intervention to ensure that corporations act more responsibly.'[26]

Job upheaval is like the threat of hanging in the morning: it concentrates the mind wonderfully. MSIs can have the same effect. Once people become aware of the hazards, they view working conditions differently. Ask how many people would be willing to take a highly paid job that carries a ten per cent risk of permanent injury, and only a few daredevils will sign up. Make that a low-paid job, and only the truly desperate will apply.

Now ask how many employed people would take a ten per cent cut in salary in exchange for job security and an extra day every week to play or spend with loved ones, and you'll be surprised how many would jump at the chance. Fifty per cent of respondents said 'Yes!' when Harvard Professor Juliet Schor asked that question in a public-opinion poll.[27] When people start thinking about their values, most see money as a means to an end rather than an end in itself.

MSIs present Canadian and U.S. societies with those very questions: would you knowingly and willingly risk your health for a job if you didn't absolutely love it? Would you prefer a chance at riches, or would you be happy enough to have some security and time for pleasure? More and more, people in North America seem to be embracing a central truth: all any of us really has in this world is time, and the health that lets us enjoy it. We can sell some of our time, but that doesn't mean that we're willing to risk our health.

# Notes

INTRODUCTION

1 Alberta Occupational Health and Safety, *Occupational Repetitive Strain Injuries*, February 1992.
2 Vernon L. Mogensen, *Office Politics: Computers, Labor, and the Fight for Safety and Health* (New Brunswick: Rutgers University Press 1996).
3 Allard E. Dembe, *Occupation and Disease: How Social Factors Affect the Conception of Work-Related Disorders* (New Haven: Yale University Press 1996).
4 Linda H. Morse, MD, MPH, and Lynn J. Hinds, MSN, 'Women and Ergonomics,' *Occupational Medicine, State of the Art Reviews* 8, no. 4 (October–December 1993).
5 Harry Braverman, *Labor and Monopoly Capital: The Degradation of Work in the Twentieth Century* (New York and London: Monthly Review Press 1974).

CHAPTER 1   MSIs, RSIs, and the Workplace

1 Vivian Smith, 'Repetitive Strain Injury Can Get Really Awful Really Fast,'*Media* 1, no. 2 (July 1994).
2 Bernardino Ramazzini, *De Morbis Artificum* [Diseases of Workers, 1713] (New York and London: Hafner Publishing 1964).
3 Dembe, *Occupation and Disease*, 38–9.
4 Gary Stix, 'Handful of Pain,' *Scientific American* (May 1991).

5 CTDNews, compilation of U.S. Bureau of Labor Statistics data, February 1996.
6 Workers' Compensation Board of British Columbia, 'Draft Ergonomics Regulations and: Statement of Context; Draft Code of Practice; Proposed Implementation Strategy,' issued by the Secretariat for Regulation Review, Board of Governors, 1993–4.
7 State of Washington Department of Labor, Division of Labor and Industries, 'Fitting the Job to the Worker: An Ergonomics Program Guideline,' 1994.
8 AFL–CIO, Ergonomic Action Alert flyer, from website http://www.aflcio.org/ergo/flyer.gif.
9 Barbara Silverstein, interview, 25 April 1996.
10 Dembe, *Occupation and Disease*, 167–8.
11 Curt Suplee, 'House to Vote on Barring RSI Rules,' *Washington Post*, 11 July 1996.
12 Visit to WHSA, November 1995.
13 Many reports, including California Ergonomics Draft Standard and CTDNews; Stuart Silverstein, 'State Unveils Scaled-Back Standards for Ergonomics, Health,' *Los Angeles Times*, 2 December 1995; and 'A Neglected Mountain of Pain: Washington Should Act Quickly on Repetitive Stress Injuries,' editorial, *Los Angeles Times*, 9 April 1996.
14 Workers' Compensation Board of British Columbia, *Occupational Health and Safety Regulations*, General Conditions (Core Requirements), sections 4.46 through 4.53, effective 15 April 1998.
15 Lynn Buekert, Lois Weininger, Janice Peterson, and Karen Webber, *Repetitive Strain Injuries in the Workplace* (Vancouver: Women and Work Research and Education Society 1991).
16 'RSI in the Workplace: Recognition and Prevention,' Conference presentation, Edmonton, 5–6 February 1993.
17 Ibid.
18 Ibid.
19 Penney Kome, telephone interview for a 'Woman's View' column, which appeared as 'Nowhere to Turn,' *Calgary Herald*, 19 April 1992.
20 Don Couch, 'Breaking Point,' *OH&S Canada* 4, no. 2 (March–April 1988).
21 Tee L. Guidotti, 'Occupational Repetitive Strain Injury,' *American Family Physician* (February 1992).

22 S.F. Ho and H.S. Lee, 'An Investigation into Complaints of Wrist Pain and Swelling among Workers at a Factory Manufacturing Motors for Refrigerators,' *Singapore Medical Journal* (June 1994).
23 Joseph D'Avanzo, 'Methods of Ergonomic Exposure Assessment: Validity and Limitations,' presentation, Fourth Managing Ergonomics Conference: Science and Policy Issues, Cincinnati, 17–20 June 1997.
24 George Botic, in *Proceedings of the Alberta Occupational Health Society Conference on Musculoskeletal Injury Prevention*, 6 November 1992.
25 Janeen Gartner, telephone interview, 3 November 1997.
26 Louise Rogers, conference presentation, in *Proceedings of the AOHS Conference on MSI Prevention*, 6 November 1992.
27 Nicole Vezina, Daniel Tierney, and Karen Messing, 'When Is Light Work Heavy? Components of the Physical Workload of Sewing Machine Operators Working at Piecework Rates,' *Applied Ergonomics* 23, no. 4 (1992).
28 Stephanie Overman, 'Cutting the High Cost of Piecework,' *Occupational Health and Safety* (October 1995).
29 Kate Pocock, 'Torture Chamber,' *Canadian Grocer* (February 1994).
30 Al McKinnon, in *Proceedings of the AOHS Conference on MSI Prevention*, 6 November 1992.
31 Ilene Stones and Wendy King, 'Office Overload: The Hidden Health Hazards of Modern Office Work,' *OH&S Canada* 7, no. 1 (January–February 1991).
32 Anthony Seaton, MD, 'Repetitive Strain Injury: Examine Working Practices,' letter to the editor, *British Medical Journal* (22 January 1994).
33 Ilene Stones, interview, November 1995.
34 Couch, 'Breaking Point.'
35 Thomas J. Armstrong, Barbara A. Silverstein, Peter Buckle, Lawrence J. Fine, Mats Hagberg, Bengt Johnsson, Asa Kilborn, Ilkka A.A. Kluorinka, Gisela Sjogard, and Eira R.A. Viikari-Juntura, 'A Conceptual Model for Work-Related Neck and Upper-Limb Musculoskeletal Disorders,' *Scandinavian Journal of Work, Environment and Health* 19, no. 3 (1993).
36 Brendan Adams, MD, interview, 15 April 1996.
37 Guidotti, 'Occupational Repetitive Strain Injury.'
38 Ontario Workplace Health and Safety Agency, *Musculoskeletal Injuries Prevention Program: Participant's Manual*, July 1992.

39 *Financial Post* report on CCOHS study of job design, 9 March 1996.
40 D.S. Chatterjee, 'Workplace Upper Limb Disorders: A Prospective Study with Intervention,' *Occupational Medicine* 42, no. 3 (1992).
41 Emil Pascarelli, MD, and Deborah Quilter, *Repetitive Strain Injury: A Computer User's Guide* (New York: John Wiley & Sons 1994).
42 *U.S. Equal Employment Opportunity Commission v. Rockwell International Corp.*, no. 95-3824 (ND IL).
43 Mahbub ul Haq, principal author, *Human Development Report 1995*, United Nations, 1995.

CHAPTER 2   Why There Are No RSIs in the Land of Oz

1 Bruce Hocking, MB, 'Epidemiological Aspects of "Repetition Strain Injury" in Telecom Australia,' *Medical Journal of Australia* 147 (7 September 1987).
2 Barbara Silverstein, interview, 25 April 1996.
3 Leslie G. Cleland, MD, 'RSI: A Model of Social Iatrogenesis,' *Medical Journal of Australia* 147 (7 September 1987).
4 Nortin M. Hadler, MD, 'Cumulative Trauma Disorders: An Iatrogenic Concept,' *Journal of Occupational Medicine* 32, no. 1 (January 1990).
5 Graham D. Wright, MB, 'The Failure of the "RSI" Concept,' *Medical Journal of Australia* 147 (7 September 1987).
6 Damian C. Ireland, MB, 'Repetitive Strain Injury,' *Australian Family Physician* 15, no. 4 (April 1986).
7 Janice Reid, Christine Ewan, and Eva Lowy, 'Pilgrimage of Pain: The Illness Experiences of Women with Repetition Strain Injury and the Search for Credibility,' *Social Science and Medicine* 32, no. 5 (1991).
8 Andrew Hopkins, 'The Social Recognition of Repetition Strain Injuries: An Australian/American Comparison,' *Social Science and Medicine* 30, no. 3 (1990).
9 Edward Byrne, 'RSI Revisited,' editorial, *Medical Journal of Australia*, 16 March 1992.

CHAPTER 3   Waiting Rooms

1 Thomas R. Hales, MD, and Patricia K. Bertsche, MPH, RN, 'Management of Upper Extremity Cumulative Trauma Disorders,' *AAOHN Journal* 40, no. 3 (March 1992).

2  Michael Lax, 'Occupational Diseases: Addressing the Problem of Under-Diagnosis,' *New Solutions* (Summer 1996).
3  Eileen Nechas and Denise Foley, *Unequal Treatment: What You Don't Know About How Women Are Mistreated by the Medical Community* (New York: Simon & Schuster 1994).
4  Brendan Adams, MD, 'Repetitive Strain Injury in the Workplace,' speech before the Alberta Occupational Health Society, May 1995.
5  Dwayne van Eerd, presentation at 'Get a Grip' workshop, Women in Media Conference, sponsored by the Canadian Association of Journalists, Toronto, November 1995.
6  Pascarelli and Quilter, *Repetitive Strain Injury*, 95. Quilter has followed this successful first book with a second, *The Repetitive Strain Injury Recovery Book* (New York: Walker and Company 1998).
7  Philip E. Higgs, MD, and Susan E. Mackinnon, MD, 'Repetitive Motion Injuries,' *Annual Review of Medicine* 46 (1995): 1–16.
8  Pascarelli and Quilter, *Repetitive Strain Injury*, 42.
9  Higgs and Mackinnon, 'Repetitive Motion Injuries,' 1–16.
10  Kathy Kilbourn, conference presentation, 'RSI in the Workplace,' Edmonton, 5–6 February 1993.
11  Peter Edgelow, 'Neurovascular Consequences of Cumulative Trauma Disorders Affecting the Thoracic Outlet: A Patient-Centered Treatment Approach,' in *Physical Therapy of the Shoulder*, ed. Robert Donatelli (New York: Churchill-Livingstone Press 1997).
12  van Eerd, presentation at 'Get a Grip' workshop.
13  Timothy Jameson, DC, 'Carpal Tunnel Syndrome: Explanation and Prevention,' *GuitarBase* (September 1996).
14  Vern Lappi, MD, 'Diagnosis of Carpal Tunnel Syndrome,' conference presentation, 'RSI in the Workplace,' Edmonton, 5–6 February 1993.
15  Dembe, *Occupation and Disease*, 73.
16  Lappi, 'Diagnosis of Carpal Tunnel Syndrome.'
17  J. Steven Moore, MD, 'Carpal Tunnel Syndrome,' *Occupational Medicine: State of the Art Reviews* 7, no. 4 (October–December 1992).
18  James H. Clay, OrthoDoc website, http://www.danke.com/Orthodoc (updated June 1997).
19  Higgs and Mackinnon, 'Repetitive Motion Industries,' 1–16.
20  Susan E. Mackinnon, MD, 'Patient Education for Cumulative Trauma Disorder,' reprinted from the *Journal of Hand Surgery* (September 1994); available from Washington University School of

Medicine, Campus Box 8508, 4444 Forest Park Ave., St Louis, MO, USA 63108–2259.
21 Dean Louis, MD, 'The Challenge from a Hand Surgeon's Perspective,' presentation, Fourth Managing Ergonomics Conference, Cincinnati, 17 June 1997.
22 Gary Franklin, MD, discussion following 'Case Definition and Diagnostic Criteria' panel, Fourth Managing Ergonomics Conference, Cincinnati, 17 June 1997.
23 Karen Strauss, 'Reflex Sympathetic Dystrophy Fact Sheet,' 1997, from the Reflex Sympathetic Dystrophy Network, 280 Riverside Drive, New York, NY 10025; fax (212) 666–6722, website http://www.rsdnet.org/.
24 M. Stanton-Hicks, W. Janig, S. Hassenbusch, J.D. Haddoz, R. Boas, and P. Wilson, 'Reflex Sympathetic Dystrophy: Changing Concepts and Taxonomy,' *Pain* 63, no. 1 (October 1995): 127–33.
25 Ken Hoelscher, MD, e-mail interview, March 1996.
26 American Osteopathic Association, 142 E. Ontario St., Chicago, IL USA 60061; toll-free 1–800–621–1773, ext. 7401; website http://www.am-osteo-assn.org/.
27 Office of Scientific and Health Communications, 'Questions and Answers about Fibromyalgia,' National Institute of Arthritis and Musculoskeletal and Skin Diseases, a component of the [U.S.] National Institutes of Health, October 1995; website http://www.nih.gov/niams/health/fibrofs.htm.
28 Jennifer Pring, Director, Educational Services for the Canadian Arthritis Society, Alberta and NWT; interview, March 1996.
29 Pfizer Canada, 'Arthritis Is the Leading Cause of Long-Term Disability,' advertisement, *Globe and Mail*, 22 February 1996.
30 Eula Bingham, MD, 'Assessing the Scope of Occupational Disease: A Candle in the Darkness,' editorial, *American Journal of Industrial Medicine* 16, no. 4 (1989): 345–6.
31 John Van Beek, telephone interview and correspondence, September 1996.
32 Association of Environmental and Occupational Health Clinics, website http://152.3.65.120/oem/aoec.htm.
33 Mark D. Gilbert, MD, Heather Tick, MD, and Dwayne Van Eerd, MSc, 'RSI: What Is It, and What Are We Doing About It?' *Canadian Journal of Rehabilitation* 10, no. 1 (1997).

34 Zoe Kessler, presentation at 'Get a Grip' workshop, Women in Media Conference, Toronto, November 1995.
35 Nancy N. Byl, Michael M. Merzenich, Steven Cheung, Purvis Bedenbaugh, Srikantan S. Nagarajan, and William M. Jenkins, 'A Primate Model for Studying Focal Dystonia and Repetitive Strain Injury: Effects on the Primary Somatosensory Cortex.' *Physical Therapy* 77, no. 3 (March 1997).

CHAPTER 4  Bigger Than a Breadbox?

1 Frank J. Testin, Research Officer, Alberta Occupational Health and Safety, to Beverly Levis, Research Officer; memorandum, 28 January 1993.
2 Renzo Bertolini and Andrew Drewczynski, *Repetition Motion Injuries* (RMI) (Hamilton, ON: Canadian Centre for Occupational Health and Safety 1990).
3 M. Martin, 'Occupational Injuries, Illness Affect 1 in 13 Canadians,' *Canadian Medical Association Journal* 153 (1995): 1782–3.
4 Statistics Canada Time-Series Matrixes H501775, H501782, and H501787, from CanSim at the Statistics Canada website, http://www.statcan.ca/.
5 U.S. Bureau of Labor Statistics report, Spring 1996.
6 Linda Rosenstock, MD, conference overview address, Managing Ergonomics Conference, Cincinnati, 17 June 1997.
7 AFL–CIO, *Stop the Pain! Repetitive Strain Injuries: An AFL–CIO Background Report*, April 1997; available from AFL–CIO, 815 16th St. NW, Washington, DC USA 20006, tel. (202) 637-5000.
8 William Weber, U.S. Bureau of Labor Statistics, telephone interview, 8 January 1998.
9 Quoted by Deborah Branscum, 'When It Hurts to Hug,' *MacWorld* (October 1989).
10 Susan K. Gauvey, 'CTDs,' from the Venables website, http://www.venable.com
11 George Erich Brogmus and Richard Markois, 'Cumulative Trauma Disorders of the Upper Extremities: The Magnitude of the Problem in U.S. Industry,' *CTDNews* 1, no. 10 (October 1992).
12 Eric Frumin, discussion, 'Size and Scope' panel, Managing Ergonomics Conference, Cincinnati, 17 June 1997.

13 'Feeling the Pain of Work? You May Have a CTD,' special report for CTD sufferers, CTD*News* (May 1993).
14 AFL–CIO, *Stop the Pain!*
15 Susan Bourette, 'Injuries Drop, But Costs Rise,' *Globe and Mail*, 25 April 1996.
16 Greg Hart, interview, April 1996.
17 Discussion from the floor at 'RSI in the Workplace' conference, Edmonton, February 1993.
18 Laura Ramsay, 'Repetitive Strain Injury a Workplace Pain,' *Financial Post*, 22 October 1992.
19 Roberta Carson, 'Reducing Cumulative Trauma Disorders,' *AAOHN Journal* 42, no. 6 (June 1994).
20 Michele Galen et al., 'Repetitive Stress: The Pain Has Just Begun,' *Business Week*, 13 July 1992.
21 'Crunch Time at the Keyboard,' *Training and Development Journal*, briefs section (April 1989).
22 Marilyn Maxim, 'A Gripping Tale,' *Benefits Canada* (February 1992).
23 'Feeling the Pain of Work? You May Have a CTD,' CTD*News* (May 1993).
24 'Average CTD Comp Case Nears $4,000,' CTD*News* (June 1996).
25 Statistics Canada, 'National Population Health Survey: Cycle 2,' The Daily Press Release, 29 May 1998. From website http://www.statcan.ca/Daily/English/980529/d980529.htm.
26 'Worker Disability Problems Rising in Industrialized Countries: Solutions Sought in Washington, DC Conference,' ILO press release, 19 May 1998.
27 Steve Mantis, telephone interview, 21 May 1998.

CHAPTER 5   Compensation? But You Don't Look Disabled

1 Penney Kome, telephone interview for article that appeared as 'Hurting Hands' in *Canadian Living* (May 1995).
2 Penney Kome, telephone interview for article that appeared as 'RSI Backlash' in *This Magazine* (May–June 1996).
3 Barbara Silverstein, interview, 25 April 1996.
4 Greg Hart, interview, May 1996.
5 Dennis Guest, *The Emergence of Social Security in Canada* (Vancouver: University of British Columbia Press 1980), 39–40. Emphasis added.

Notes to pages 75–82   215

6 Alberta Federation of Labour, *Securing Just Compensation: A How-to Manual to Assist Workers in Winning Compensation*, rev. ed., 1991.
7 Ibid.
8 Guest, *Emergence of Social Security*, 44.
9 Workers' Compensation Board of British Columbia, 'Medical and Legal Issues Related to the Recognition of Occupational Disease: A Briefing Paper,' from website http://www.wcb.bc.ca.
10 Workers' Compensation Board of British Columbia, 'Comparison of Occupational Health and Safety Statutes: A Briefing Paper' (18 March 1997); from website http://www.wcb.bc.ca/new/briefing/brief7.htm#anchor220705.
11 Ontario Workers' Compensation Board, 'Minister Witmer Announces Major Overhaul of Workers' Compensation in Ontario,' news release, 26 November 1996, from website http://149.174.222.20/wcb/wcb.nsf/Public/WCBReformpg2.
12 Ontario Ministry of Labour, 'Workers' Compensation Reform Clears Another Hurdle,' media release, 9 October 1997.
13 Ontario Ministry of Labour, 'Preventing Illness & Injury: A Better Health and Safety System for Ontario Workplaces,' January 1998.
14 George Botic, Canadian Auto Workers' union presentation to the Manitoba Federation of Labour Health and Safety Enforcement Meeting, Winnipeg, 18 February 1996.
15 AFL–CIO, 'Stand UP for Worker Protections,' fact sheet from the 'Stand UP for America's Working Families' campaign, AFL–CIO Department of Occupational Safety and Health, tel. (202) 637–5210, website http://www.aflcio.org/standup/suprotec.html.
16 Diane Pravikoff, MSN, RN, and Joyce A. Simonowitz, MSN, RN, 'Cumulative Trauma Disorders: Developing a Framework for Prevention,' *AAOHN Journal* 42, no. 4 (April 1994).
17 Ibid.
18 Paul Holyoke and Robert Elgie, 'The Many Crises in Workers' Compensation,' *OH&S Canada* (May–June 1993).
19 Rona Maynard, 'The Pain Threshold,' *Canadian Business* (February 1993).
20 Susan Bourette, 'Workplace Injuries Drop, But Costs Rise,' *Globe and Mail*, 25 April 1996.
21 Carson, 'Reducing Cumulative Trauma Disorders.'
22 Canadian Labour Congress, 'CLC-Policy: Briefing Note on Workers'

Compensation,' from website http://www.clc-ctc.ca/policy/ (undated).
23 Steve Chase, 'Sick Leave Soars,' *Calgary Sunday Sun*, 28 April 1996.
24 Janet McFarland, 'The War Against WCB Fraud,' *Globe and Mail*, 19 November 1995.
25 Terry Lavendar, 'Strains and Sprains More Likely at the Beginning of the Week, Study Says,' *At the Agency Newsletter* 5, no. 3 (May 1995).
26 Greg Hart, interview, May 1996.
27 Barbara Silverstein, interview, 25 April 1996.
28 Mary Fricker, 'Widespread Fraud: Bogus Claim,' *Sonoma County Press Democrat*, 8 December 1997.
29 Mary Fricker, 'State Audits Find Insurers Routinely Unfair to Workers,' *Sonoma County Press Democrat*, 8 December 1997.
30 AFL–CIO, 'Facts on Workers' Compensation and RSIs,' from http://www.aflcio.org/safety/factson.htm.
31 Ibid.
32 American Academy of Orthopaedic Surgeons, 'Legislation Enacted in 1997: Workers' Compensation,' 22 January 1998; from http://www.aaos.org/wordhtml./stateleg/en_work.htm.
33 Oregon Health and Safety Resource, 'Oregon OSHA Renews Its Efforts to Reduce Ergonomics Injuries' (Winter 1997), from website http://www.cbs.state.or.us/external/osha/admin/resource/spring97.htm#ergo1.
34 Pennsylvania Chamber Workers Compensation Network, 'How You Can Get Involved: Business Groups/Associations/Businesses,' from website http://204.127.237.2/wcinvolv.htm (updated 1996). Emphasis added.
35 The Hon. Elizabeth Witmer, MPP, *Hansard*, 26 November 1996. Emphasis added.
36 The Hon. Richard Patten, MPP, *Hansard*, 24 April 1997.
37 Business Council of New York State, 'The Cost of Comp: A Business Council Briefing Paper on Workers' Compensation and New York State's Economy,' March 1996, from website http://www.bcnys.org/member/lg/workers_comp/wc_costs.htm.
38 Amy Clipp, *Job Damaged People: How to Survive and Change the Workers' Comp System*, a joint project of the Environmental Health Network and the Louisiana Injured Workers' Union; 152 pages, $19.95

U.S.; from EHN Books, P.O. Box 16267, Chesapeake, VA, USA 23328–6267, tel. (757) 424–1162, fax (747) 424–1517.
39 New York Committee on Occupational Health and Safety, flyer, 1996, used with permission.
40 Steven Babitsky, telephone interview, February 1996; 'Backlash!' *Workers' Compensation Monthly* 16, no. 1 (January 1996).
41 Canadian Labour Congress, 'CLC-Policy: Briefing Note on Workers' Compensation.'
42 *Canadian Auto Workers Health, Safety and Environment Newsletter* 5, no. 6 (June 1997).
43 Clipp, *Job Damaged People*, 7.
44 Conference presentation, 'RSI in the Workplace: Recognition and Prevention,' Edmonton, 5–6 February 1993. Following Ramonalee's presentation, Susan Ruffo (then chair of the Alberta Workers' Health Centre) mentioned that every year about a quarter of all workers at that particular poultry-packing plant file WCB claims. Poultry packing is high-risk for MSIs. But Ramonalee's second claim might have been viewed as suspicious because repeated claims are often perceived as possible indicators of fraud.
45 Monica Zurowski, 'Woman's Work Injury Snagged in WCB Files,' *Calgary Herald*, 12 March 1993.
46 Clipp, *Job Damaged People*, 20.
47 Conference presentations and discussion from the floor, 'RSI in the Workplace: Recognition and Prevention,' Edmonton, 5–6 February 1993.
48 Ibid.
49 Canadian Press, 'Salaries Soar for Alberta WCB Execs: Pay Surpasses Other Provinces,' *Globe and Mail*, 12 July 1997.
50 Kelly Eby, Alberta Workers' Compensation Board, telephone interview, 17 July 1997.
51 Kevin Flaherty, telephone interview, July 1997.
52 Conference presentation, 'RSI in the Workplace: Recognition and Prevention,' Edmonton, 5–6 February 1993.
53 Freeda Steinberg, 'The Law of Workers' Compensation as It Applies to Hand Injuries,' *Occupational Medicine: State of the Art Reviews* 4, no. 3 (July–September 1989).
54 Brooke E. Smith, 'The Bogus Epidemic: Repetitive Stress Injuries and

218   Notes to pages 96–103

Computer Keyboarding,' from website http://www.hal-pc.org/journal/01repet.html.
55  Center for Office Technology, 'Report on State Treatment of Cumulative Trauma Disorders,' 10 December 1996.
56  Ontario Workers' Compensation Board, 'Executive Summary' of the comprehensive review by the Hon. Cam Jackson, Minister Without Portfolio, 1996.
57  Ontario Workers' Compensation Board, 'Fact Sheet II: Refocusing the System as a Workplace Insurance Plan,' from website http://149.174.222.20/wcb/wcb.nsf/Public/WCBReformpg7.
58  Marie-Anne Roiseux, personal e-mail correspondence, 22 August 1996.
59  New Brunswick Department of Economic Development and Tourism, 'New Brunswick: Call Centre Capital of North America,' from website http://www.cybersmith.net/econ-dev/solution/.
60  Workers' Compensation Board of British Columbia, 'Ergonomics (MSI) Requirements,' sections 4.46 to 4.53, *Occupational Health and Safety Regulations*, effective 15 April 1998.

CHAPTER 6   So Sue Me

1   Rory O'Neil, 'RSI Risk: They've Known for Years,' *The Journalist* (October–November 1994).
2   Compaq Computer, *Creating a Comfortable Work Environment*, September 1991.
3   Peter Reeve, 'Woman Awarded Record Damages of £186,000 for RSI,' *The Guardian*, 24 January 1997.
4   Gauvey, 'CTDs,' from the Venables website.
5   Edward Felsenthal, 'Out of Hand: An Epidemic or a Fad?' *Wall Street Journal*, 14 July 1994.
6   Gauvey, 'CTDs,' from the Venables website.
7   *Davis v. NCR*. 'A CTD Sufferer Finally Wins Product Liability Lawsuit,' CTD*News* (August 1996).
8   *Johnson v. Norfolk & Western Railway Co.*, Virginia Supreme Court, summarized in CTD*News*, 'Legal Briefs' (August 1996).
9   'A CTD Sufferer Finally Wins.'
10  Cerisse Anderson, 'Punitive Award Is Barred in Suit Over Keyboards,' *New York Law Journal*, 17 January 1997.

11 'New Evidence Overturns $5 Million Keyboard Verdict: Judge Orders New Trial in *Geressy* v. *Digital Equipment Corp.*,' CTD*News* (June 1997).
12 Reynolds Holding, 'RSI Suit May Finally Catch Up with Apple,' *San Francisco Chronicle*, 19 January 1997.
13 'Memorandum in Support of Plaintiffs' Motion for Various Relief Concerning the Deposition of Dr R. Bruce Hocking,' *Repetitive Stress Injury Litigation Reporter* (March 1996).
14 *Zarecki* v. *National Passenger Railroad Corp., d/b/a Amtrak*, No. 95–1075 (ND IL).
15 *Ciesielski* v. *Norfolk & Western Railway Co.*, No. 93–1554 (WD PA). (From RSI *Ligitation Reporter*, March 1996.)
16 *U.S. Equal Employment Opportunity Commission* v. *Rockwell International Corp. et al.*, No. 95–3824 (ND IL).

CHAPTER 7   Staring at the Screen: Computer Workstations

1 CTD*News*, 'Introduction,' *The North American Ergonomic Resources Guide*, 1995–6 ed.
2 Public Service Alliance of Canada, *PSAC Manual on VTDs and Your Health*, 1992. Emphasis in original.
3 Jim Dixon, 'Making the Office User-Friendly,' *OH&S Canada* 5, no. 1 (1989).
4 Richard Wolkomir, 'When the Work You Do Ends Up Costing You an Arm and a Leg,' *Smithsonian Magazine* (June 1994).
5 Rochelle Sharpe, 'Ergonomic Devices Can Be Hazardous to Your Health,' *Wall Street Journal*, 9 April 1996 (Work Week section).
6 Bob Morency, office ergonomist, SOREHAND mailing list discussion, 4 September 1996 (among other dates).
7 'Prevention,' e-mail newsletter published by VDT Solution, St Charles, IL, USA.
8 Pascarelli and Quilter, *Repetitive Strain Injury*, 167.
9 ErgoWeb, 'Summary Checklist for Computer (VDT) Workstation Risk,' from website http://ergoweb.com/; Pasacarelli and Quilter, *Repetitive Strain Injury*, 40.
10 Mackinnon, 'Patient Education for Cumulative Trauma Disorder.'
11 Cindy Moser, 'Computer Tracking: New-Age Stress,' editorial, *OH&S Canada* 7, no. 1 (January–February 1991).

12 Rochelle Sharpe, 'Typing as a Sport?,' *Wall Street Journal*, 9 April 1996 (Work Week section).
13 Branscum, 'When It Hurts to Hug.'
14 Barbara Silverstein, 'Can In-Plant Exercise Control Musculoskeletal Symptoms?' *Journal of Occupational Medicine* 30, no. 12 (1988).
15 K. Lee, N. Swanson, S. Sauter, R. Wicktron, A. Waikar, and M. Mangum, 'A Review of Physical Exercises Recommended for VDT Operators,' *Applied Ergonomics* 23, no. 6 (1992): 387–408.
16 Stephanie Brown, *The Hand Book* (New York: Ergonome Inc. 1992).
17 Choon-Nam Ong, 'Musculoskeletal Disorders in Operators of Visual Display Terminals,' *World Health Forum* 15, no. 2 (1994): 161–4.
18 Stones and King, 'Office Overload: The Hidden Health Hazards of Modern Office Work.'
19 Public Service Alliance of Canada, *PSAC Manual on VTDs and Your Health*.
20 Mary Gooderham, 'High-Tech RSI Aid Creates New Problem,' *Globe and Mail*, 14 September 1995.
21 For more up-to-date information on voice recognition, visit Susan Fulton's website, http://www.outloud.com
22 Cyril A. Wantland, Subhas C. Gupta, MD, and Scott A. Klein, 'Safety Considerations for Current and Future VR Applications,' from website http://www.webmed.com/.
23 Paul Linden, *Compute in Comfort* (Englewood Cliffs, NJ: Prentice Hall 1995).

CHAPTER 8  Fitting the Jobs to the Workers

1 Moira Farr, 'Work That Wounds and How to Heal It,' *Canadian Business* (December 1991).
2 *CTDNews*, 'An Ergonomics Honor Roll: Case Studies of Results-Oriented Programs,' special report, 1996.
3 Kathleen Buckheit, OHN, interview, 19 June 1997.
4 *CTDNews*, 'An Ergonomics Honor Roll.'
5 ErgoWeb, http://ergoweb.com/.
6 Ibid.
7 Cathy Walker, interview, 17 July 1997.
8 At various times, OSHA has been prohibited from publishing or disseminating any portion of its draft ergonomics standard. Still, these

signal risk factors and many other guidelines may be viewed on the Internet at ErgoWeb (http://ergoweb.com/) or at http://ctdnews.com/aal.html.
9 Workers' Compensation Board of British Columbia, 'Draft Ergonomics Regulations.'
10 John Van Beek, telephone interview, September 1996.
11 Hank Lick, PhD, interview, 17 June 1997.
12 Jack MacDonald, 'Many Workers Won't Admit They Need Help,' *Calgary Herald*, 4 November 1995.
13 Frank Hamade, 'Repetitive Strain Injuries at Work: Are You At Risk?,' *Alliance* 8, no. 4 (Winter 1995–6).
14 Linda H. Morse, MD, and Lynn J. Hinds, MSN, 'Women and Ergonomics,' *Occupational Medicine: State of the Art Reviews* 8, no. 4 (October–December 1993).
15 Barbara A. Silverstein, Lawrence J. Fine, and Thomas J. Armstrong, 'Hand–Wrist Cumulative Trauma Disorders in Industry,' *British Journal of Industrial Medicine* 43 (1986): 779–84.
16 State of Washington Department of Labor, 'Fitting the Job to the Worker: An Ergonomics Program Guideline.'
17 James McCauley, interview, 19 June 1997.
18 Hales and Bertsche, 'Management of Upper Extremity Cumulative Trauma Disorders.'
19 Barbara Silverstein, PhD, 'Upper Limb and Low Back Musculoskeletal Disorders: State and National Estimates Based on Workers' Compensation Accepted Claims,' presentation, Fourth Managing Ergonomics Conference, Cincinnati, 17–20 June 1997.
20 Women and Work Research and Education Society, 'Repetitive Strain Injury Questionnaire – Report for British Columbia,' paper distributed at the RSI in the Workplace conference, Edmonton, February 1993.
21 'Working Conditions Study: Big 3 Getting Worse,' *Canadian Auto Workers Health, Safety and Environment Newsletter* (July 1996).
22 'Pain at Work,' *Globe and Mail* (Business section, Management Briefs), 18 June 1996.
23 ErgoWeb, http://ergoweb.com/.
24 Ontario Workplace Health and Safety Agency, *Musculoskeletal Injuries Prevention Program: Participant's Manual*, July 1992; entries for 'Clerical & Related Professions,' 'Processing, Assembling & Repairing,' 'Construction,' and 'Manufacturing.'

25 Workers' Compensation Board of British Columbia, 'Draft Ergonomics Regulations and: Statement of Context; Draft Code of Practice; Proposed Implementation Strategy.'
26 Ibid.; State of Washington Department of Labor, Division of Labor and Industries, 'Fitting the Job to the Worker: An Ergonomics Program Guideline.'
27 See CTD*News*, *North American Ergonomic Resources Guide, 1995–6*, 99–108. 'The draft pre-proposals materials are intended to facilitate informed discussion as the [OSHA] develops an Ergonomic Protection Standard. These materials are not being promulgated as a rule or a standard. They do not constitute either a Notice of Proposed Rulemaking or a source of official OSHA guidance on ergonomics.'
28 Barbara Silverstein, interview, 25 April 1996.
29 Gail Sater, 'Intervention and Controls,' presentation at the Fourth Managing Ergonomics Conference, Cincinnati, 18 June 1997.
30 Lida Orta-Anes, PhD, interview, 17 June 1997.
31 United Food and Commercial Workers, 'Ergonomics Programs: The Way to Prevent Cumulative Trauma Disorders (CTDs),' n.d.; available from the UFCW International Union, 1775 K Street NW, Washington, DC 2006–1598, USA.
32 Manitoba Workplace Safety and Health, 'Ergonomics at Work: One Company's Success Story,' *WorkSafe* (February 1996).
33 CTD*News*, 'What Works to Cut CTD Risk, Improve Job Productivity?' *North American Ergonomic Resources Guide, 1995–6*.
34 Workers' Compensation Board of British Columbia, 'Comparison of Occupational Health and Safety Statutes: A Briefing Paper.'
35 Andrew Card, Jr, 'Managing Ergonomics in the '90s,' keynote address to the American Automobile Manufacturers' Association, 28 February 1995.
36 Canadian Auto Workers Health and Safety Fact Sheet, 'Overuse Injuries,' Occupational Health and Safety Issue No. 6, n.d.
37 Tony Horwitz, 'Nine to Nowhere: These Six Growth Jobs Are Dull, Dead-End, Sometimes Dangerous,' *Wall Street Journal*, 1 December 1994.
38 'Burns Bathroom Breaks Policy Must Go,' *Labour News* (October 1994).
39 David C. Alexander, 'The Economics of Ergonomics,' presentation, Fourth Managing Ergonomics Conference, Cincinnati, 18 June 1997.

40 CTD*News*, 'An Ergonomics Honor Roll: Case Studies of Results-Oriented Programs.'
41 Moser, 'Computer Tracking: New-Age Stress.'
42 Farr, 'Work That Wounds.'
43 Hank Lick, PhD, interview, 17 June 1997.
44 Don Couch, 'The Economics of Ergonomics,' case studies, *OH&S Canada* 6, no. 5 (1989).
45 Ibid.
46 Barbara Silverstein, interview, 25 April 1996.

CHAPTER 9  Beyond Grieving

1 AFL–CIO, 'Ergonomics Alert,' from website http://www.aflcio.org/ergo/flyer.gif.
2 Peg Semanario, 'Overview,' presentation to the Fourth Managing Ergonomics Conference, Cincinnati, 17 June 1997.
3 U.S. Bureau of Labor Statistics, 1994.
4 Jim Selby, 'Unionized Workers Better Educated, Better Paid and More Secure,' *Labour News* [a publication of the Alberta Federation of Labour] (April 1996); based on Diane Galarneau, *Perspectives on Labour and Income*, Statistics Canada, 1996.
5 Conference presentation, 'RSI in the Workplace: Recognition and Prevention,' Edmonton, 5–6 February 1993.
6 Dembe, *Occupation and Disease*, 85–6.
7 'CAW Statement of Principles: Health and Safety,' from Canadian Auto Workers' website, http://www.web.net/caw/policy/cawhs.html (updated 11 November 1996).
8 New Brunswick Health, Safety and Compensation Commission, minister's speech, 26 April 1996.
9 AFL–CIO, 'Stand UP for America: Safer Workplaces,' Ergonomic Action flyer, from website, http://www.aflcio.org/standup/suprotec.htm.
10 Alberta Workers' Health Centre, fact sheet and brochure, 1995.
11 Public Service Alliance of Canada, 'On April 28, Spot the Hazards,' *Alliance* (Spring 1996).
12 Brian Kohler, 'April 28 – A Day of Mourning,' Communications, Energy and Paperworkers' Union of Canada fact sheet, 1995.

13 Public Service Alliance of Canada, 'On April 28, Spot the Hazards.'
14 Frank Hamade, telephone interview, 27 August 1996.
15 Treasury Board Manual, *Personnel Management Component, Occupational Safety and Health*, chapter 5–5, 'A Guide on Video Display Terminals' (30 June 1993).
16 Cathy Walker, interview, 17 July 1997.
17 United Food and Commercial Workers and Bakery, Confectionery and Tobacco Workers' Union, 'Look Into My Eyes. They Have a Story to Tell!' advertisement, *Calgary Herald*, 30 April 1997.
18 Jeff Adams, 'Workers Say Cargill Production Demanding,' *Calgary Herald*, 13 July 1997.
19 Mogensen, *Office Politics*, 42.
20 'Wire Service Guild – Negotiated Improvements in AP Contracts Since the 1980s,' information sheet, n.d., from WSGWeb, http://www.wsg.org/e222/wsginfo/contract/progress.htm#health-safety.
21 Gail Lem, discussion at Women in Media conference, Toronto, November 1995; Institute of Work and Health website, http://www.iwh.on.ca.
22 Confédération des syndicats nationaux (CSN), 'Les Employés syndiqués à la CSN obtiennent la semaine de quatre jours,' media release, 5 June 1996.
23 Office and Professional Employees' International Union, 'The President's Corner,' from OPEIU On-Line, http://www.opeiu.org (updated 8 July 1996).
24 Retail–Wholesale Canada/United Steelworkers of America, 'Workers' Compensation Then and Now,' *RW Banner* no. 2 (July 1995).
25 Alberta Workers' Health and Safety Centre, Edmonton, 1996.
26 Steven Greenhouse, 'Labor's Labors Not Lost,' *New York Times*, 12 May 1996.
27 AFL–CIO, 'Organizing, Mobilizing and Re-Awakening Hope: Media Advisory for Labor Day,' 29 August 1996, from website http://www.aflcio.org/publ/press96/pr08261.htm.
28 Mogensen, *Office Politics*, 39.
29 U.S. Bureau of Labor, *Report on Conditions of Women and Child Wage-Earners in the United States*, vol. 10, 1991; as cited at the Labor Quotes website, http://igc.apc.org/laborquotes.

30 Lin Lim, principal author, *More and Better Jobs for Women*, International Labour Organization report, July 1996.
31 International Labour Organization, 'Women Swell Ranks of Working Poor, Says ILO,' ILO press release, 30 July 1996.
32 Susan Crean, *Grace Hartman: A Woman for Her Time* (Vancouver: New Star Books 1995).
33 Penney Kome, *Women of Influence: Canadian Women and Politics* (Toronto: Doubleday Canada 1985).
34 Laurell Ritchie, 'Why Are So Many Women Unorganized?' in *Union Sisters: Women in the Labour Movement*, ed. Linda Briskin and Lynda Yanz (Toronto: Women's Press 1983).
35 Canada NewsWire, Canadian Auto Workers press release, 18 June 1996.
36 Louise Laporte, 'Reaching the Level Playing Field,' *Alliance* 8, no. 4 (Winter 1995–6).
37 Judy Darcy, 'Afterword,' in Crean, *Grace Hartman*.
38 Canadian Labour Congress, *Women's Work: A Report by the Canadian Labour Congress*, ed. Winnie Ng, 8 March 1997.
39 AFL–CIO, 'Stand UP for America: Safer Workplaces,' flyer.
40 Environmental Research Foundation, 'A Political Opportunity,' *Rachel's Environment & Health Weekly*, no. 507 (15 August 1996); available from P.O. Box 5036, Annapolis, MD 21403, USA; fax (419) 263-8944; e-mail erf@rachel.clark.net.
41 Interfaith Committee for Worker Justice, 'Action Alert – Southern Poultry Workers Need Your Help,' Internet e-mail appeal, November 1996.
42 L. Paltrineri and Don Mackle, 'Poultry Workers Strike in N. Carolina,' *The Militant* 60, no. 31, 9 September 1996.
43 National Congress of Employees, 'Historical Background' and 'NCE Benefits to Members,' from website, http://www.nce.org/ (indexed on Yahoo's Labor home page).
44 'Workers of America: Providing a Voice for the Concerns of the American Worker' and 'Summary of Major Worker Issues,' from Workers of America website, http://www.vais.net/~bweiner/woa/html#Why WoA Was Created (indexed on Yahoo's Union home page).
45 National Employee Rights Institute, 'NERI,' from website http://www.disgruntled.com/neriad.html.

46 Working Today website at http://www.workingtoday.org/; Jobs With Justice website at http://www.igc.apc.org/jwj/; Americans with Work-Related Injuries website at http://web2.airmail.net/mskapar2; America's MedCheck Organization website at http://www.telepath.com/medcheck/workcomp.html.
47 James C. Robinson, 'Worker Responses to Workplace Hazards,' *Journal of Health Politics, Policy and Law* 12, no. 4 (Winter 1987).
48 Buzz Hargrove, '1996 Labour Day Message from Canadian Auto Workers President Buzz Hargrove,' 2 September 1996.
49 Susan E. Bourette, 'Workplace Injuries Drop, But Costs Rise,' *Globe and Mail*, 25 April 1996.
50 Humberto Marquez, 'Unions Want More "Democratic" Globalisation,' InterPress Third World News Agency, distributed worldwide via the APC networks, 13 August 1996.
51 Canadian Labour Congress, 'Global Solidarity: A Trade Union International Agenda,' policy statement, n.d.

CHAPTER 10  Legislation in Other Jurisdictions

1 Lesley Meall, 'The VDU: Bad for Your Health?' *Accountancy* (June 1992).
2 Secretary of State for Employment (U.K.), Health and Safety Executive, 'Approved Code of Practice: Management of Health and Safety at Work Regulations, 1992.' Emphasis in original.
3 Secretary of State for Employment (U.K.), Health and Safety Executive, 'HSE Launches National Workplace Health and Safety Week and New Guidance Showing Employers How to Prevent RSI,' media release #071–717 6918, 17 October 1994.
4 Trade Union Congress (U.K.), publications for the 1996 'Don't Suffer in Silence' campaign against upper-limb disorders, from website http://congress96.tuc.org.uk/scripts/browse.exe?155&0.
5 Ibid., campaign briefing #12, August 1996.
6 K.C. Parsons, 'Ergonomics of the Physical Environment,' *Applied Ergonomics* 26, no. 4 (1995).
7 European Union Occupational Health and Safety, 'General Framework for Action by the European Commission in the Field of Safety,

Hygiene and Health Protection at Work, 1994–2000,' 4.1.1a (November 1993).
8 'Les Risques du travail sur ordinateur,' from FranceWeb, http://www.franceweb.fr/sosinformatique/. Emphasis in original.
9 Suzanne Kolare, 'An Unsuitable Job for Muscles and Bones,' in *Strategies for Prevention of Musculoskeletal Disorders*, National Institute of Occupational Health (Sweden), February 1993.
10 Peter Tillhammar, 'The Method That Makes Muscular Load Visible,' in ibid.
11 Kolare, 'An Unsuitable Job for Muscles and Bones.'
12 Ministry of Labour (Japan) Ordinance No. 59, 1992 [replaces Industrial Safety and Health Law No. 47, 1972]: 'Principles Pertaining to the Measures to be Carried Out by Business Proprietors for Facilitating the Establishment of Comfortable Working Environments'; kindly supplied by the Japan Industrial Safety and Health Association (JISHA) of the Ministry of Labour.
13 Ibid.
14 Labour Standards Bureau (Japan), 'Guidelines to Occupational Health in VDT Operation,' Notice No. 705, issued 20 December 1985; and Occupational Health Division, Occupational Health and Safety Department (Japan), 'Official Information of Commentary on Points to Note in the Implementation of the Guidelines to Occupational Health in VDT Operation,' 17 March 1986; both kindly supplied by the Japan Industrial Safety and Health Association (JISHA) of the Ministry of Labour.
15 Labour Standards Bureau (Japan), ibid.
16 WorkSafe Australia, 'National Code of Practice for the Prevention of Occupational Overuse Syndrome,' from the National Occupational Health and Safety Commission [NOHSC: 2013 (1994)].

CHAPTER 11  The Battle over Legislation

1 David Felinski, welcoming speech, Fourth Managing Ergonomics Conference, Cincinnati, 17 June 1997. The entire proceedings of the conference are available from ErgoWeb at http://ergoweb.com/.
2 *Congressional Record*, 25 April 1996, H3901.

3 Curt Suplee, 'House to Vote on Barring RSI Rules,' *Washington Post*, 11 July 1996.
4 California Occupational Safety and Health Standards Board, 'California Proposed Ergonomics Regulation,' 1996.
5 'A Neglected Mountain of Pain: Washington Should Act Quickly on Repetitive Stress Injuries,' editorial, *Los Angeles Times*, 9 April 1996.
6 Raju Narisetti, 'Battle Lines Form,' *Wall Street Journal*, 1 November 1994 (Work Week section).
7 J. Steven Moore, MD, 'Disorders of the Muscle–Tendon Units of the Distal Upper Extremity,' Fourth Managing Ergonomics Conference.
8 Alfred Franzblau, MD, 'A Cross-Sectional Study of the Relationship Between Repetitive Work and Upper Extremity Musculoskeletal Disorders,' Fourth Managing Ergonomics Conference.
9 William Marras, PhD, 'A Prospective Validation of the LMM Low Back Disorder Risk Model,' Fourth Managing Ergonomics Conference.
10 Don Chaffin, PhD, 'Role of Body Motion Models,' Fourth Managing Ergonomics Conference.
11 Kurt T. Hegman, MD, 'Application of the Strain Index: An Advance in Exposure Assessment and Analysis,' Fourth Managing Ergonomics Conference.
12 Jeffrey A. Arons, MD, Jeffrey C. Salomon, MD, and Marvin S. Arons, MD, letter, *Journal of Hand Surgery* (January 1997), 163–5.
13 Susan E. Mackinnon, MD and Christine B. Novak, PT, MS, 'Clinical Perspective: Repetitive Strain in the Workplace,' *Journal of Hand Surgery* (January 1997), 2–18.
14 Nortin M. Hadler, MD, 'Clinical Perspective: Repetitive Upper-Extremity Motions in the Workplace Are Not Hazardous,' *Journal of Hand Surgery* (January 1997), 19–29.
15 Arons, Salomon, and Arons, letter, *Journal of Hand Surgery*.
16 National Economic Research Associates, 'Costs and Benefits of Ergonomics Regulations,' released by the American Trucking Association, 4 December 1996, from website http://www.trucking.org/press/ergo.html.
17 Citizens for Reliable and Safe Highways, 'Fatigue: The Facts,' 1996; from http://www.trucksafety.org/facts/htm or crashsf@aol.com or 1 (800) CRASH-12.

18 Canadians for Responsible and Safe Highways, 'CRASH Calls for Action Plan on Truck Safety,' media release, 21 February 1997, from http://www.web.net/~crash/press.html or crash@web.net or (613) 860–0529 or CRASH, Box 1042, Station B, Ottawa, ON K1P 5R1.
19 Jeffrey Kluger, 'Risky Business,' *Discover* (May 1996), 44–7.
20 Adam M. Finkel, 'Who's Exaggerating?' *Discover* (May 1996), 48–54.
21 Ibid.
22 Terrence H. Murphy, 'OSHA and Ergonomics,' presentation, Fourth Managing Ergonomics Conference, Cincinnati, 19 June 1997.
23 Joseph D'Avanzo, 'Methods of Ergonomic Exposure Assessment: Validity and Limitations,' presentation and discussion, Fourth Managing Ergonomics Conference, Cincinnati, 19 June 1997.

CHAPTER 12  By the Fingernails

1 Brad Evenson, 'Shorter Jobs and Longer Layoff Signs of Times,' Southam News, *Calgary Herald*, 11 May 1996.
2 Louis Uchitelle and N.R. Kleinfield, 'On the Battlefields of Business, Millions of Casualties,' *New York Times*, 3 March 1996.
3 Tony Horwitz, 'Nine to Nowhere,' *Wall Street Journal*, 1 December 1994.
4 Victor Keegan, 'A World Without Bosses – Or Workers,' *The Guardian* (London); reprinted in *Globe and Mail*, 24 August 1996.
5 Bruce Little, 'Forty Hours: A Workweek on the Skids,' *Globe and Mail*, 2 September 1996.
6 Canadian Labour Congress, 'New Strategies for Health, Safety and Workers' Compensation – Final Report,' November 1996.
7 Braverman, *Labor and Monopoly Capital*, 82–3.
8 David Heyman, 'Lost Time Due to Injury a Cargill Issue,' *Calgary Herald*, 21 July 1997.
9 Paul R. Krugman, 'Stay on Their Backs,' *New York Times Magazine*, 4 February 1996.
10 Jeremy Rifkin, *The End of Work* (New York: Putnam Books 1995).
11 Bruce O'Hara, *Working Harder Isn't Working* (Vancouver: New Star Books 1993), 60–1.
12 'The Trouble with Men,' *The Economist*; reprinted in *Globe and Mail*, 5 October 1996.

13 Evenson, 'Shorter Jobs and Longer Layoffs.'
14 Angus Reid, *Shakedown: How the New Economy Is Changing Our Lives* (Toronto: Doubleday Canada 1996), 271–2.
15 Dembe, *Occupation and Disease*, 257.
16 Swedish National Institute for Occupational Health, 'Time Pressure, a Pain in the Neck' in *Strategies for Prevention of Musculoskeletal Disorders*, National Institute of Occupational Health (Sweden), February 1993.
17 Greg Hart, interview, April 1996.
18 Maria Benktzon, 'Designing for Our Future Selves: The Swedish Experience,' *Applied Ergonomics* 24, no. 1 (February 1993).
19 Karen Unland, 'Canadians Pay for Busy Lives,' *Globe and Mail*, 19 July 1997.
20 Reid, *Shakedown*, 190, 208.
21 O'Hara, *Working Harder Isn't Working*, 98.
22 Rifkin, *The End of Work*, 28.
23 Ibid., 26–7.
24 Michael Valpy, 'The New, Value-Added Canadians,' *Globe and Mail*, 26 October 1996 [interview with Neil Nevitte about his book, *The Decline of Deference* (Peterborough, ON: Broadview Press 1996)]. .
25 'Corporate Irresponsibility: There Ought to Be Some Laws' [report of a survey conducted by EDK Associates of New York], *Rachel's Environment and Health Weekly*, no. 507 (15 August 1996); copies available from the Preamble Center for Public Policy, 1737 21st Street NW, Washington, DC 20009, tel. (202) 265–3263 or from ftp.std.com/periodicals/rachel.
26 Ibid.
27 Jesse Hahnel, 'Is Time Really Money?' *Dollars and Sense Magazine* (January–February 1998).

# Selected Bibliography

BOOKS

AFL–CIO. 'Stop the Pain! Repetitive Strain Injuries: An AFL–CIO Background Report,' April 1997

Alberta Federation of Labour. *Securing Just Compensation: A How-to Manual to Assist Workers in Winnipeg Compensation.* Edmonton: Alberta Federation of Labour 1991

Alberta Occupational Health Society. *Proceedings of the Conference on Musculoskeletal Injury Prevention,* 6 November 1992

Anderson, Bob. *Stretching at Your Computer or Desk.* Shelter Publications 1997

Braverman, Harry. *Labor and Monopoly Capital: The Degradation of Work in the Twentieth Century.* Monthly Review Press 1974

Briskin, Linda, and Lynda Yanz. *Union Sisters: Women in the Labour Movement.* Toronto: Women's Press 1983

Brown, Stephanie. *The HAND Book.* New York: Ergonome 1993

Buekert, Lynn, Lois Weninger, Janice Peterson, and Karon Webber. 'Repetitive Strain Injuries in the Workplace.' Women and Work Society & the Alberta Workers' Health Centre 1991

Butler, Sharon J. *Conquering Carpal Tunnel Syndrome and Other Repetitive Strain Injuries.* Oakland, CA: New Harbinger Publications 1996

Canadian Labour Congress. *Women's Work: A Report.* Ottawa: CLC 1997

Center for Office Technology. 'Office Ergonomics Management Program,' 1993

Clipp, Amy. *Job Damaged People: How to Survive and Change the Workers' Comp System*, a joint project of the Environmental Health Network and the Louisiana Injured Workers' Union, 1993

*CTDNews, The North American Ergonomic Resources Guide.* Horsham, PA: LRP Publications 1996–7

Dembe, Allard E. *Occupation and Disease: How Social Fractors Affect the Conception of Work-Related Disorders.* New Haven, CT: Yale University Press 1996

Donatelli, Robert. *Physical Therapy of the Shoulder.* New York: Churchill-Livingstone 1997

Guest, Dennis. *The Emergence of Social Security in Canada.* Vancouver: University of British Columbia Press 1980

Kolare, Suzanne. *Stragegies for Prevention of Musculoskeletal Disorders.* Stockholm: Swedish National Institute of Occupational Health 1993

Menzies, Heather. *Whose Brave New World?* Toronto: Between the Lines 1996

Mogensen, Vernon L. *Office Politics: Computers, Labor and the Fight for Safety and Health.* New Brunswick, NJ: Rutgers University Press 1996

Nechas, Eileen, and Denise Foley. *Unequal Treatment: What You Don't Know about How Women Are Mistreated by the Medical Community.* New York: Simon & Schuster 1994

O'Hara, Bruce. *Working Harder Isn't Working: How We Can Save the Environment, the Economy and Our Sanity by Working Less and Enjoying Life More.* Vancouver: New Star Books 1993

Pascarelli, Emil, MD, and Deborah Quilter. *Repetitive Strain Injuries: A Computer User's Guide.* New York: John Wiley & Sons 1994

Public Service Alliance of Canada. *PSAC Manual on VDTS and Your Health.* Rev. ed. Ottawa: PSAC 1992

Quilter, Deborah. *The Repetitive Strain Injury Recovery Book.* New York: Walker and Company 1994

Ramazzini, Bernardino. *De Morbis Artificum* [Diseases of Workers, 1713]. Translated by Wilmer Cave Wright. Riverside, NJ: Hafner Publishing Co. 1964

Rifkin, Jeremy. *The End of Work: The Decline of the Global Labor Force and the Dawn of the Post-Market Era.* New York: C.P. Putnam's Sons 1995

## Selected Bibliography

GOVERNMENT PUBLICATIONS

Alberta Occupational Health & Safety. 'Occupational Repetitive Strain Injuries,' 1992

European Union Occupational Health and Safety. 'General Framework for Action by the European Commission in the Field of Safety, Hygiene and Health Protection at Work 1994–2000,' November 1993

Health and Safety Executive, Secretary of State for Employment, U.K. 'Approved Code of Practice: Management of Health and Safety at Work Regulations, 1992.'

Ministry of Labour, Japan. 'Ministry of Labor Ordinance No. 19, 1992, Replaces Industrial Safety and Health Law No. 57, 1972: Principles Pertaining to the Measures to Be Carried Out by Business Proprietors for Facilitating the Establishment of Comfortable Working Environments,' 1992

National Institute for Occupational Safety and Health. *Cumulative Trauma Disorders in the Workplace: Bibliography.* U.S. Department of Health and Human Services Publication No 95-119, 1995

Treasury Board Manual. 'Personnel Management Component, Occupational Safety and Health, Chapter 5-5: A Guide on Video Display Terminals (30-06-93)'

Workers' Compensation Board of British Columbia. 'Comparison of Occupational Health and Safety Statutes: A Briefing Paper,' 1997

– Draft Ergonomics Regulations and: Statement of Context; Draft Code of Practice; Proposed Implementation Strategy,' Secretariat for Regulation Review, British Columbia Board of Governors, 1995

– 'Medical and Legal Issues Related to the Recognition of Occupational Disease: A Briefing Paper,' 1997

Workplace Health and Safety Agency. *Musculoskeletal Injuries Prevention Program: A Head-to-Toe Ergonomics Training Program, Participant's Manual. Clerical and Related Industries.* Toronto: Publications Ontario 1992

– *Musculoskeletal Injuries Prevention Program. Construction.* Toronto: Publications Ontario 1992

– *Musculoskeletal Injuries Prevention Program. Manufacturing.* Toronto: Publications Ontario 1992

– *Musculoskeletal Injuries Prevention Program. Processing, Assembling & Repairing*. Toronto: Publications Ontario 1992

MEDICAL AND SCIENTIFIC JOURNALS

Bingham, E. 'Assessing the Scope of Occupational Disease: A Candle in the Darkness.' *American Journal of Industrial Medicine* 16, no. 4 (1989): 345–6
Chatterjeee, D.S. MD. 'Workplace Upper Limb Disorders: A Prospective Study with Intervention.' *Occupational Medicine* 42, no. 3 (1992)
Couch, Don. 'The Economics of Ergonomics.' *OH&S Canada* 6, no. 5 (1989)
Gilbert, Mark D., MD, Heather Tick, MD, Dwayne Van Eerd, MSC. 'RSI: What Is It, and What Are We Doing About It?' *Canadian Journal of Rehabilitation* 10, no. 1 (1997)
Guidotti, Tee, MD. 'Occupational Repetitive Strain Injury.' *American Family Physician* (February 1992)
Hales, Thomas R., MD, and Patricia K. Bertsche, MPH, RN. 'Managememt of Upper Extremity Cumulative Trauma Disorders.' *AAOHN Journal* 40, no. 3 (March 1992)
Lee, K., N. Swanson, S. Sauter, R. Wicktron, A. Waikar, and M. Manguma. 'A Review of Physical Exercises recommended for VDT Operators.' *Applied Ergonomics* 23, no. 6 (1992)
Mackinnon, Susan E., MD, Christine B. Novak, PT. 'Clinical Perspective: Repetitive Strain in the Workplace.' *Journal of Hand Surgery* (January 1997)
Moore, J. Steven, MD. 'Carpal Tunnel Syndrome.' *Occupational Medicine: State of the Art Reviews* 7, no. 4 (October–December 1992)
Morse, Linda H., MD, and Lynn J. Hinds, MSN. 'Women and Ergonomics.' *Occupational Medicine: State of the Art Reviews* 8, no. 4 (October–December 1993)
Moser, Cindy. 'Computer Tracking: New-Age stress.' *OH&S Canada* 7, no. 1 (January–February 1991)
Silverstein, Barbara A. 'Can In-Plant Exercise Control Musculoskeletal Symptoms?' *Journal of Occupational Medicine* 30, no. 12 (1988)
Silverstein, Barbara A., Lawrence J. Fine, Thomas J. Armstrong. 'Hand–Wrist Cumulative Trauma Disorders in Industry.' *British Journal of Industrial Medicine* 43 (1986): 779–84

Stones, Ilene, and Wendy King. 'Office Overload: The Hidden Health Hazards of Modern Office Work,' *OH&S Canada* 7, no. 1 (January–February 1991)

MAGAZINES AND NEWSPAPERS

Branscum, Deborah. 'When It Hurts to Hug.' *MacWorld* (October 1989)

Chase, Steve. 'Sick Leave Soars.' *Calgary Sun*, 28 April 1996

CTDNews. 'Ergonomics Becoming Bigger Labor Contract Issue.' (August 1996)

Editorial, 'A Neglected Mountin of Pain: Washington Should Act Quickly on repetitive Strain Injuries,' *Los Angeles Times*, 1996

Farr, Moira. 'Work That Wounds and How to Cure It.' *Canadian Business* (December 1991)

Finkel, Adam. 'Who's Exaggerating?' *Discover Magazine* (May 1996)

Galen, Michele. 'Repetitive Stress: The Pain Has Just Begun,' *Business Week*, 13 July 1992

Greenhouse, Steven. 'Labor's Labors Not Lost.' *New York Times*, 12 May 1996

Horwitz, Tony. 'Nine to Nowhere: These Six Growth Jobs Are Dull, Dead-End, Sometimes Dangerous,' *Wall Street Journal*, 1 December 1994

Kluger, Jeffrey. 'Risky Business.' *Discover Magazine* (May 1996)

McFarland, Janet. 'The War Against WCB Fraud.' *Globe and Mail*, 19 November 1995

Maynard, Rona. 'The Pain Threshold.' *Canadian Business* (February 1993)

Mittlestaedt, Martin. 'Injured Workers to Lose $15.2 Billion Under Ontario Plan.' *Globe and Mail*, 4 July 1996

Public Service Alliance of Canada. 'National Day of Mourning.' *Alliance* (Spring 1996)

Selby, Jim. 'Unionized Workers Better Educated, Better Paid and More Secure,' *Labour News* (Spring 1996)

Wolkimir, Richard. 'When the Work You Do Ends Up Costing You and Arm and a Leg,' *Smithsonian Magazine* (June 1994)

Zurowski, Monica. 'Woman's Work Injury Snagged in WCB Files.' *Calgary Herald*, 12 March 1993

## Selected Bibliography

WEBSITES

AFL–CIO, *http://www.aflcio.org/*
Canadian Auto Workers, *http://www.web.net/caw/*
Canadian Labour Congress, *http://www.clc-ctc.ca/*
Clay, James. 'Orthodoc' (massage therapy, theory, practice and muscle diagrams) *http://danke.com/Orthodoc/*
*Disgruntled* e-zine, *http://www.disgruntled.com/*
ErgoWeb, *http://www.ergoweb.com*
'FindADoc,' *http://www.engr.unl.edu/eeshop/findadoc.html* or by e-mail: message to *mpaul@engrs.un.edu* with subject, 'Findadoc'
Graps, Amara. 'Amara's RSI Page,' *http://www.amara.com/aboutme/rsi.html*
Marxhausen, Paul. 'Computer RSI Page,' *http://www.engr.unl.edu/eeshop/rsi.html*
National Institute for Occupational Safety and Health (U.S.), *http://www.cdc.gov/niosh/homepage.html*
Occupational Safety and Health Administration (U.S.), *http://www.osha-slc.gov/SLTC/Ergonomics/*
Wallach, Dan [now maintained by Wright, Scott]. 'Typing Injury FAQ,' available at: *http://www.tifaq.com*
Workers' Compensation Board of British Columbia, with background papers and links to other provincial WCB sites, *http://www.wcb.bc.ca/*

# Appendix

CANADIAN INJURED WORKERS' GROUPS

Canadian Council of Rehabilitation and Work
20 King Street West, 9th floor
Toronto, ON M5H 1C4
Ph. (416) 974-5575, fax (416) 974-5577
TTY (416) 974-2636
http://www.ccrw.org/index.html

Canadian Injured Workers' Alliance/L'Alliance canadienne des victimes d'accidents et de maladies du travail
CP/Box 3678
237 rue Camelot Street
Thunder Bay, ON P7A 4B2
Ph. (807) 345-3429, fax (807) 345-7086
http://indie.ca/ciwa/
*Co-ordinates information among eighty-three local injured workers' groups.*

Cape Breton Injured Workers' Group
PO Box 204
New Waterford, NS B1H 4N9
Ph. (902) 567-6768, fax (902) 849-0652

Injured Workers' Alliance of Nova Scotia
11 Westwood Drive
Dartmouth, NS B2X 1Y3

Injured Workers Human Rights Group of BC
Lower Mainland Chapter
Vancouver Island Chapter
PO Box 274
Gabriola Island, BC V0R 1X0
Ph. (250) 390-9107, fax (250) 390-9108

London Occupational Safety & Health Information Service
222-424 Wellington Street
London, ON N6A 3P3
Ph. (519) 433-4156 or (519) 433-2887
http://www.execulink.com/~losh/enhanced/Emain.htm
*Hosts an RSI support group.*

Ontario Network of Injured Workers' Groups
President: Carl Krevar
10 Wise Crescent
Hamilton, ON L8T 2L5

St. John Labour Community Services Inc.
560 Main Street
Saint John, NB E2K 1J5
Ph. (506) 635-0391 or (506) 634-7099
http://www.sjlabour@nbnet.nb.ca

Thunder Bay & District Injured Workers Support Group
Victoriaville Mall, Suite 109-B
125 S Syndicate Ave
Thunder Bay, ON P7E 6H8
Ph. (807) 622-8897, fax (807) 622-7869

Toronto Injured Workers' Group
c/o Injured Workers' Consultants
815 Danforth Avenue, Suite 411
Toronto, ON M4J 1L2
Ph. (416) 461-2411
http://www.injuredworkers.org

WRIST
Worker's (Repetitive) Injury Support Team, Woodstock and Area
http://www.geocities.com/CapitolHill/Lobby/2614/
*Site lists twenty groups in the Ontario Network.*

# Index

Action Canada, 158
acupressure, 44
acupuncture, 44
Adams, Brendan, 18, 32, 53, 68, 116, 191
Aetna Insurance, 186
affirmative action programs, 156
AFL–CIO: Ergonomic Action Alert flyer, 142; ergonomics regulation and, xiii; on injuries and deaths, 145–6; on rate of MSIs, 8, 64–5, 66–7; on responsibility for MSIs, 86, 151–2; 'Stand Up' campaign, 78; on standards, 181; 'Stop the Pain!' campaign, 146–7; women workers and, 157–8
age, MSIs and, 95
Alberta: privatization in, 155–6, 158; Women and Work in, 151; workers' compensation in, 77, 93–4; workers' health centres in, 151
Alberta Federation of Labour, 75, 137, 143, 156
Alberta Labour, 62–3, 77
Alberta Union of Public Employees (AUPE), 156
Alberta Workers' Health Centre, 146
Alberta's Common Front, 158
Alexander, David C., 138–9
Alexander technique, 36, 45
American Academy of Orthopaedic Surgeons, 87
American Association for Hand Surgery, 182–3
American National Standards Institute (ANSI), 101
American Society for Surgery of the Hand, 182
American Trucking Association, 181, 183–4
*Americans with Disabilities Act*, 104
Americans with Work-Related Injuries, 161
*Amtrak, Zarecki v.*, 104
analgesics, 35. *See also* pain
Anderson, Betty, 103–4

Apple Computers, Inc., 103
Armstrong, Thomas, 18, 124
Aronsson, Gunnar, 199
arthritis, 55–6
Asia Pacific Economic Community (APEC), 163
assembly-line work, xv, 10, 16, 65, 137, 193
Associated Press, 148
Association for Repetitive Motion Syndromes (ARMS), 58
Association of Environmental and Occupational Health Clinics, 57
Association of Occupational and Environmental Clinics, 39
athletes, 12
AT&T (San Diego), 119
Australia: compared with United States, 28–9; Code of Practice, 174–5; MSIs in, 23–9
auto industry, 13–14, 65, 127, 135–7, 149–50, 193
automation, 194–8, 203
avocations, MSIs and, 21

B., Karen, 12
Babbage, Charles, xv
Babitsky, Steven, 91
back pain, 170, 187
Barnes, Stephanie Schoenfeldt, 58
Beedle, Warren, 139
Bell Canada, 113
benefits, disability, 191, 192
Benktzon, Maria, 199
Berkman, Mark, 183–4
Bertsche, Pat, 30–1, 126

Bingham, Eula, 57
Biswanger, Kathy, 93
blacklisting, of injured workers, 89
blue-collar workers, 29, 65, 137
Board of Certification for Professional Ergonomists (BCPE), 120
bodywork, 44–5, 108
Bonilla, Henry, 179, 180
Botic, George, 14, 78
Bourette, Susan, 163
brachial plexus, 48
Braverman, Harry, 192–3
breaks from work, xv, 20–1, 172–3; exercise, 35, 132; in newspaper offices, 113–14. *See also* rest
breathing, deep, 44–5
Breckenridge, Joan, 12, 33, 149
British Columbia, 150: ergonomics regulations in, 129–30, 131; legislation in, 10; occupational health and safety committees, 135; Occupational Health and Safety Regulations, 10; workers' compensation in, 7, 98–9, 120–1
British Safety Council, 167
British Telecom, 167
Broadway, Michael, 193
Brogmus, George Erich, 66
Brown, Stephanie, 114–15
Buckheit, Kathleen, 119
Buerkert, Lynn, 151
Business Council of New York State (BCNYS), 88, 89
Butler, Sharon, 44
buyouts, 90

C., Shauna, 72–3, 74
California: ergonomics regulations in, 9–10, 177–8, 180–1; Federation of Labor, 9–10, 180–1; workers' compensation in, 78–80
Canada: ergonomics regulation in, 175–6; medical care in, 41, 59–60
Canada Mortgage and Housing Corporation (CMHC), 71–2
Canadian Arthritis Society, 55
Canadian Auto Workers (CAW), 149–50; on compensation rates, 91, 127; on hazard information, 144–5; on hours of work, 148; on prevention, 136–7; on rate of MSIs, 13–14; strike, 202
Canadian Centre for Occupational Health and Safety (CCOHS), 21, 63, 64
Canadian Injured Workers' Alliance, 76
Canadian Labour Congress, 82, 91, 98, 156, 163–4, 192
Canadian Standards Association, 147
Canadian Union of Public Employees, 157
Canadians for Responsible and Safe Highways (CRASH), 184
capital assets vs. human assets, 192
Card, Andrew H. Jr., 135
carpal tunnel syndrome (CTS): causes, 42–3; diagnosis, 40, 45–52; lawsuits, 104; in meat packing, 178; personal factors in, 95–6; surgery for, 49, 126; tests for, 32; in the U.K., 167; in women, xiii, 12, 67
Carr, Vic, 144
cashiers, supermarket, 15–16, 194
casinos, 16, 149, 150
Center for Office Technology, 96
Centre for Injury and Disease Response, 40
Chaffin, Don, 182
chairs, design of, 107–8
Chase, Tashlyn, 187
Chavez-Thompson, Linda, 151
Chemical, Energy and Paperworkers Union, 146
children: care of, 156, 200; living on street, 198; MSIs in, 109
chiropractors, 42–3
*Cieselski v. Norfolk & Western Railway Co.*, 104
Citizens for Reliable and Safe Highways (CRASH), 184
Cleland, Leslie G., 25
clerical workers, xiii, 6, 112, 115
Cleveland Clinic, 47
Clinic of Injury and Disease Response (CIDR), 58
Clipp, Amy, 161
collective bargaining, xiii, 148–9
Communications Workers of America (CWA), 94
Communications Workers Union (U.K.), 167
Compaq, 102
Complex Regional Pain Syndrome (CRPS), 51

Computer RSI Page, 59
computers, 102: injuries from use, xiv; musculoskeletal injuries (MSIs) and, 105–17; work deskilling and, 193–4; workers, 169–70; workstations, 105–17. *See also* automation; electronics assembly; keyboarding; visual display terminals
Conference on Managing Ergonomics in the 1990s, 67, 122, 127, 177–8, 181–2, 186–9
contract employment, 191, 192
corporations: hours of work and, 204; treatment of workers, 159
costs: of MSIs, 67–70; of occupational injuries, 81–2, 140; of workers' compensation, 74, 81–2; workforce, 68
Couch, Don, 12, 17–18
crafts, MSIs and, 95
cramps: muscle, 37–8; telegrapher's, 6; writer's, 5–6
Crean, Susan, 154, 157
CTD*News*: 'best practices,' 139; on costs of MSIs, 68–9; on ergonomics, 119, 134; on keyboards, 110; on rate of MSIs, 63–4, 66, 185; on regulations, 138
Cuddy Foods, 118, 139
cumulative trauma disorder (CTD), 4, 7; diagnosis, 79
cutbacks, government: to healthcare, 69; to public service, 197

Darcy, Judy, 157
D'Avanzo, Joseph, 13, 187

Davies, Frank, 166
Davis, Eula, 103
Day, Stockwell, 82–3
deaths, in workplace, 145–7
Dembe, Allard E., 6, 47, 198
dentists, 65
deskilling, of work, 6, 193–4
diagnosis: of CTS, 45–8; of MSIs, 25, 33, 38
disability: benefits, 191, 192; legislation regarding, 201–2; long-term, 82–3; pensions, 89–92; risk of MSIs and, 109
diseases, occupational, 76
*Disgruntled* e-zine, 160–1
Dixon, Jim, 107
Dole, Elizabeth, 178
*Dole v. Dow*, 101
*dollars and SENSE* magazine, 87
domestic violence, 156
domestic work, 196–7
Don't Suffer in Silence campaign, 166
Dorman, Peter, 69
downsizing, 190
Drewczynski, Andrew, 21, 63

ear protectors, 8
early intervention, 126
Eby, Kelly, 94
Edgelow, Peter, 41, 60
Edington, P.J., 188
electronics assembly, xv, 10, 16, 65, 137, 193. *See also* assembly-line work; computers
Elgie, Robert, 80–1

employers: fraud by, 85; worker/
  employer teams, 125
epicondylitis, 48
equipment, 20; changes to, 122–3;
  design of, 199; new, 13
ergonomics, xiii, 102–3, 118–41;
  consultants, 138; cost of programs, 138–41, 185, 186, 188;
  evaluation of, 36; legislation,
  8–9, 175–89; in product design,
  138; productivity and, 121; regulations, xiii, 9–10, 77, 99,
  120–1, 129–30, 175–89, 188–9;
  surveillance of, 30–1; in Sweden, 199; trade unions and, xiii;
  in the United Kingdom, 166;
  women and, 123–4
ergonomics-related disorder, 4
ergonomists, 53–4, 120
ErgoWeb, 120, 129
European Community, VDU regulation in, 165
European Union, labour standards in, 168–9
Ewan, Christine, 26–8
exercises, 34, 35, 53, 132

factory work, 16
fatigue, 22
Feldenkrais technique, 36, 45
Felinski, David, 177
FFD (force, frequency, duration),
  19, 121
fibromyalgia, 55
Find-a-Doc, 39
Fine, Lawrence J., 124, 171, 188
Finkel, Adam M., 185, 186

first aid, 34–5
Fisher, Doug, 93
Flaherty, Kevin, 94
Foley, Denise, 32
force, use of, 19, 121
Ford, Henry, 192
Ford Motor Company, 122, 136,
  139–40, 186
forearm supports, 110
Forest, James, 83
France, ergonomics legislation in,
  169–70
FranceWeb, 169
Franklin, Gary, 51
Franzblau, Alfred, 181
fraud, in workers' compensation,
  82–6
Fricker, Mary, 86
Frumin, Eric, 66

Gartner, Janeen, 14
Gauvey, Susan K., 65, 102–3
gender: ergonomics and, 124;
  union membership and, 143.
  *See also* women
General Motors, 14, 202
General Seating, 119, 139
*Geressy* v. *Digital Equipment Corp.*,
  103
glass ceiling, 197
globalization: effect on labour
  standards, 177–8; trade unions
  and, 163–4
*Globe and Mail*, 149
golf elbow, 48
Gonzales, Patrisia, 104
Good Jobs campaign, 161

Gooderham, Mary, 116
grocery workers, 15–16. *See also* supermarket cashiers
Guest, Dennis, 74, 75
Guidotti, Tee L., 13, 18, 57
gyromouses, 111

Hadler, Nortin M., 25, 183
Hagberg, Mats, 170
Hale, Thomas, 30–1
Hamade, Frank, 123, 147
hands: use of force, 121. *See also* carpal tunnel syndrome (CTS); SOREHAND listserv; wrist position
Hansenne, Michel, 153
Harding, Reynolds, 103
Hargrove, Buzz, 162
Hart, Greg, 54, 67, 73, 84–5, 122, 132, 199
Hartman, Grace, 155, 157
headsets, 36
health management organizations (HMOs), 60–1
health professionals, 15
healthcare: cutbacks in, 69; universal, 41, 59–60, 198. *See also* insurance
hearing loss, 8–9, 109
Hegman, Kurt, 182
Hellerwork, 44
Higgs, Philip E., 38, 40, 49
Hinds, Lynn J., xiv, 124
hobbies, MSIs and, 95
Hocking, Bruce, 24, 104
Hoelscher, Ken, 53
Holyoke, Paul, 80–1

home: housework, 12, 95, 196–7; inequality in, xiii; working from, 200
Hopkins, Andrew, 28–9
hormones, MSIs and, 32, 95
Horwitz, Tony, 190–1
hours of work, 111–12, 147–8, 202–4. *See also* overtime

IBM, 100
ice, as treatment, 34
immigrants, union membership and, 155
Industrial Conference Board, 203
Injured Workers' Day, 147
Injured Workers' Union, 77
injuries: causes, 187; costs of, 81–2, 140
Institute for Work and Health, 127
insurance: employment, 159; health, 41, 59–60; no-fault, 76
Interfaith Committee for Worker Justice, 160
International Labour Organization (ILO), 70, 153–4
Ireland, Damian, 26

Jacobsen, Patricia, 81
Jameson, Timothy, 42–3
Janeway, Barbara, 7
Japan, labour legislation in, 172–4
Japan Industrial Safety and Health Association, 172
Japanese Association of Industrial Health VDT Work Study Committee, 173
jobs: automation and, 194–6;

design of, 20–1, 187; deskilling of, 193–4; elimination of, 197; repetition in, 18–19, 121; re-skilling of, 199; rotation, 132–3; security of, 190–1, 202; variety in, 115–16
Jobs For All, 161
Jobs With Justice, 161

K., Donna Lee, 71–2, 74
K., Ramonalee, 92–3, 144
Karpowich, Linda, 156
Keegan, Victor, 191
Kentucky, workers' compensation in, 86
Kessler, Zoe, 59
keyboarding, 29, 109–10, 112, 169
Kilbom, Asa, 170
Kilbourn, Kathy, 41
kinesiologists, 14, 53, 54
King, Wendy, 16–17, 115
Kluger, Jeffrey, 185
Kohler, Brian, 146
Krugman, Paul R., 194

labour. *See* workers; workforce
Labour Standards Bureau (Japan), 173
Lantos, Tom, 65
Laporte, Louise, 157
Lappi, Vern, 46, 47
lawsuits, product liability, 100–4
lawyers, workers' compensation and, 94
Lax, Michael, 31
Leamon, Tom, 65–6
legislation, 9–10, 175–89

leisure: pain and, 187
Liberty Mutual Insurance Company, 65–6
Lick, Hank, 122, 136, 139–40
Lighten Your Load campaign, 166
lighting, 20, 107
Lim, Lin, 153
Lost Time Claims (LTC), 62–3
Louis, Dean, 50–1, 182
Louis Harris and Associates, 127–8
Lowy, Eva, 26–8
Luxton, David, 167

M., Pam, 51–3
MacIntyre, Roly, 145
MacKinnon, Al, 16
Mackinnon, Susan E., 38, 40, 49–50, 109, 112, 183
Mackle, Don, 160
magic bullets, 56
Maine, workers' compensation in, 96
Malling, Eric, 131
Manitoba: occupational health and safety committees, 135; workers' health centres in, 39, 151
Manitoba Federation of Labour, 77–8
Manitoba Workplace Safety and Health Branch, 133
Manpower Inc., 158
Manpower Temporary Help, 200–1
Mantis, Steve, 70
Markois, Richard, 66

Marras, William, 182
Marxhausen, Paul, 39, 58
massage therapy, 34–5, 36, 43–4
Maynard, Rona, 81, 82
McCauley, Jim, 126
McGuire, Geraldine, 157
McMaster University, 127, 136
measurements, anthropometric, 123–4
meat packing, 14, 64, 67, 137, 178, 186, 193; strike, 148
MedCheck Organization, 161
medical care, 41, 59–60. *See also* healthcare
menopause, as cause of MSIs, 32
Meredith, Justice William, 75, 76
Mettert, Margaret, 105
middle management, 197
Mirer, Frank, 177
Mogensen, Vernon, 148, 152
Moore, J. Steven, 48, 181
Morse, Linda H., xiv, 124
Moser, Cindy, 112–13, 139
mouse, computer, 110–11
movement training, 36, 117
MSK injuries. *See* musculoskeletal injuries (MSIs)
Murphy, Terrence H., 187
Murray, T.J., 98
muscle load, measurement of, 171
musculoskeletal injuries (MSIs): blame for, 76; causes of, 33, 34, 95–6; cost of, xiv, 67–70, 166; definition, 3–4; diagnosis, 25, 34, 38; effect of, 66–7; epidemics, 23; history, 5–7; Information Age and, 198; in Japan, 172; labour strikes and, 147–8; legislation regarding, 9–10, 175–89; media coverage, 24–5; medical management of, 125–6; personal factors, 21, 95–6; pre-employment testing for, 21; prevention, 9, 64, 87; psychogenic component, 26–7, 84–5; psychological factors, 31–2; rate of, xiv, 7–8, 13–14, 62–7; regulation of, 9–10, 175–89; research, 13, 61; responsibility for, 95–6; risk factors, 17–22; self-care, 34–7; stages of, 5, 11, 84–5; statistics, 13–14, 62–7; symptoms, 24–5; temporary workers and, 201; treatment, 32, 34, 56–60, 125–6; working conditions and, 205–6. *See also* breaks; repetitive strain injury (RSI); therapies
myofascia, 61
myofascial pain syndrome, definition of, 4
myotherapy, 44

narrow compression, 44
National Association of Manufacturers, 181
National Coalition on Ergonomics (NCE), 180, 183
National Congress of Employees (NCE), 160
National Council on Compensation Insurance, Inc., 91
National Day of Mourning, 145
National Economic Research Associates (NERA), 183

National Employee Rights Institute (NERI), 161
National Employment Lawyers' Association, 161
National Health and Safety in the Workplace Week, 166
National Institute for Occupational Safety and Health (NIOSH), 9, 64, 114, 180, 189
National Institute of Arthritis and Musculoskeletal and Skin Diseases, 55
National Occupational Health and Safety Commission (Australia), 174
National Population Health Survey, 70
NCR Corp., 103
Nechas, Eileen, 32
neck ache, 111
nerves, 48–9; compression of, 40
neuromuscular therapy, 44
Nevitte, Neil, 204
New Brunswick: worker deaths in, 145; workers' compensation in, 10, 77, 98
New York Committee on Occupational Safety and Health, 90
New York (State), workers' compensation in, 88, 89, 90
newspaper offices, 24–5, 113
Ng, Winnie, 157
no-fault insurance, 76
non-steroidal anti-inflammatories (NSAIDs), 32
*Norfolk & Western Railway Co., Cieselski v.*, 104

North American Free Trade Agreement (NAFTA), 163, 168
Northwest Territories, workers' compensation in, 77
Nova Scotia: occupational health and safety committees, 135; Women and Work in, 151; workers' compensation in, 98
Novak, Christine, 183
nurses, 15; in workplaces, 30–1
nutrition, 36

occupational diseases, 65, 76
occupational health and safety committees, 134
Occupational Health Clinics, 39
Occupational Health Clinics for Ontario Workers (OHCOWs), 57, 121, 151
occupational health nurses, 30–1
Occupational Medicine Consultation Clinic, 57
occupational overuse syndrome (OOS), 4, 23, 174–5
Occupational Safety and Health Administration (OSHA), 9, 29, 63; in California, 78–80; codes of practice, 129; ergonomics code, 188–9; ergonomics standards, 178–80; standards, 78–80
Occupational Safety and Health Standards Board (California), 180
occupational therapists, 42
Office and Professional Employees' International Union, 150

250  Index

Office Ergonomics Research Committee, 110
office workers, 10–11, 16–17, 65, 147
O'Hara, Bruce, 195, 202–3
Ohio, workers' compensation in, 86–7
O'Neil, Rory, 100
Ong, Choon-Nam, 115
Ontario: occupational health and safety committees, 135; RSI in, 9; WCB fraud in, 83; workers' compensation in, 10, 77, 78, 89, 97; workers' health centres in, 39, 151
Ontario Federation of Labour, 151
Ontario Ministry of Labour, 63
Ontario Network of Injured Workers, 147
Ontario Workplace Health and Safety Agency, 129
Oregon, workers' compensation in, 87
Orta-Anes, Lida, 133
osteoarthritis, 55
osteopaths, 54
overexertion, 63
overtime, 148, 202. *See also* hours of work

packaging materials manufacture, 133–4
pain, 10–11, 34–5, 127; back, 170, 187; chronic, 81, 97, 98; neck, 111; relievers, 56; in upper extremities, 4
Paltrineri, L., 160

Parent, Madeleine, 154
Parsons, K.C., 168
part-time workers, 191, 192; union membership and, 155; women, 157
Pascarelli, Emil, 34, 38, 43, 108, 111
Pastula, Susan, 136
Patch, Kimberly, 87
Patten, Richard, 89
pay equity, 157
Pennsylvania, workers' compensation in, 88
Perdue Poultry, 126
Pfizer Canada, 56
Phalen, George S., 47
physiatrists, 53
physical therapy, 40–2
physicians: choice of, 39; fees, 60; MSIs and, 31–3
piecework, 15, 191, 200
Pike, Ron, 93
PIMEX (Picture Mix Exposure), 171
pink-collar workers, 196. *See also* clerical workers
polarization, of society, 195
postal workers, 16
post-materialism, 204
posture, 9, 17, 19–20, 36, 53, 108, 121, 124
poultry processing, 14, 67, 118, 137, 144, 160
Pravitkoff, Diane S., 78–9
prevention. *See* musculoskeletal injuries (MSIs)
Prince Edward Island: workers'

compensation in, 77; workers' health centres in, 139, 151
privatization: of public service, 155–6, 158; of workers' compensation, 77
product liability lawsuits, 100–4
productivity: costs of, xiv; ergonomics and, 121, 140; monitoring of, 112–13
professional workers, 105–6, 150
profit motive, in medical care, 60
pronator teres muscle, 112
Prudden, Bonnie, 44
public service: cutbacks, 197; privatization, 155, 158
Public Service Alliance of Canada, 106, 115–16, 146, 147, 157

Quebec: occupational health and safety committees in, 135; workers' compensation in, 97
Quilter, Deborah, 34, 111

R., May, 11
Rabinowitz, Randy, 188
railroad workers, 104
Ramazzini, Bernardino, xiv, 5–6
recreation, MSIs and, 13, 95, 187
Red Wing Shoes, 119, 132–3, 186
Reeve, Gordon, 136
Reflex Sympathetic Network, 51
Regional Inter-American Organization of Workers (ORIT), 163
regional pain syndrome, 4, 23, 46
Reid, Angus, 197, 202
Reid, Janice, 26–8
Rempel, David, 110

repetition, in jobs, 18–19, 121
repetitive strain injury (RSI): in Australia, 23; definition of, 3–4; psychological factors, 26–7; sociological factors, 28–9; treatment of, 42; women and, 26–8. *See also* musculoskeletal injuries (MSIs)
responsibility; internal, 77–8; for MSIs, 95–6
rest: as treatment for MSIs, 32–3. *See also* breaks
Retail–Wholesale Canada, 150
rheumatoid arthritis, 55
rheumatologists, 49
Riche, Nancy, 156
Rifkin, Jeremy, 194–5, 203
risk assessment, 185
risk factors, 17–22; in computer use, 106–12; non-job-related, 187
Ritchie, Laurell, 155
Robinson, James C., 162
Rogers, Louise, 15
Rogers, Sue, 187
Roiseux, Marie-Anne, 97
Rolfing, 44
Rosenstock, Linda, 64, 180, 181
Rotolo, Jeanette, 103

Safeway, strike at, 148
Saskatchewan, occupational health and safety committees in, 135
Sater, Gail, 132–3, 139
Saunders, Sharon, 94
Savardi, David, 188

Schor, Juliet, 205
scissor-lifts, 132
secretaries, 193–4
Selby, Jim, 143, 156
self-help groups, 34–7, 58–9
Semanario, Peg, 142–3, 189
Semorile, Trina, 7
sewing, 15
sexual harassment, 154
shiatsu, 44
Silverstein, Barbara: on Australia, 25; on 'cascade model,' 18; on gender and ergonomics, 124; on MSI profile, 73; on newsrooms, 113; in NIOH, 171; on posture, 19, 108; on rate of MSIs, 8, 127; on risk factors, 187, 188–9; on Sweden and Japan, 173; on teamwork, 131–2; on 'trigger finger,' 133; on Washington state, 96; on WCB fraud, 85; on worker need to organize, 141
Simonowitz, Joyce A., 78–9
Simpson, Muriel, 101
slant boards, 110
sleep, 36, 40, 55
SOAP model, 46–7
social assistance, 197–8. *See also* welfare
social costs, of MSIs, 69
social iatrogenesis, 25
somatics, 45
SOREHAND listserv, 35, 49, 59, 60, 116
Southern Ontario Newspaper Guild (SONG), 149
speech. *See* true voice recognition

sports: medicine, 42; musculo-skeletal injuries (MSIs) and, 95
Statistics Canada, 63–4, 66–7, 70, 143–4
Steinberg, Freeda, 95
Stewart, Patrick, 45
Stones, Ilene, 16–17, 21, 113, 115
Strauss, Karen, 51
stress, in VDT workers, 115
stretching, 34, 35, 36, 38, 114, 132. *See also* breaks; exercises
strikes, labour, 147–8
stripping (therapy), 44
Suplee, Curt, 179–80
support groups, 34–7, 58–9
surgery, for CTS, 46, 49–52
Sweden: ergonomics in, 199; MSIs in, 170–1; re-skilling of jobs in, 199
Swedish National Institute of Occupational Health (NIOH), 170–1, 199
Sweeney, John J., 151
syndromes, 33

t'ai chi, 44
tasks. *See* jobs
Taqi, Ali, 70
Taylor, Paul, 149
teams, 131–2; multidisciplinary, 56–8; in workplaces, 125
Telecom Australia, 23, 24
telecommuting, 200
telegrapher's cramp, 6
telephone use, 111
temperatures, at work, 20
tendinitis, 12, 66

tennis elbow, 48
therapies, 34–6, 43–4. *See also* musculoskeletal injuries ; (MSIs): treatment
tools, 20, 121
tracking devices, 110–11
Trade Union Congress, 166–7
trade unions, 25–6, 142–58; ergonomics and, xiii; globalization and, 163–4; hours of work and, 203; membership in, 159–60; in New Brunswick, 97–8; part-time workers in, 155; public view of, 152; women in, 152–8
Trades and Labour Congress of Canada, 75
Treasury Board Manual, 147
truckers, 14, 184
true voice recognition, 116–17

unemployment, xiv, 158–9, 195–6
Union of Needletrades, Industrial and Textile Employees (UNITE), 66
unions. *See* trade unions
United Auto Workers (UAW), 133, 148, 149
United Food and Commercial Workers, 133, 144, 178
United Kingdom, health and safety standards, 165–8
United States: Bureau of Labor Statistics (BLS), 64; compared with Australia, 28–9; *Equal Employment Opportunity Commission v. Rockwell International Corp. et al.*, 104; ergonomics regulation in, 175–6, 178–80; lawsuits in, 101–4; medical care in, 60–1, 96–7; MSIs in, 178–80; occupational health in, 198–9; unions in, 156; workers' compensation in, 78–80, 87
United Steelworkers of America, 150
Universal Product Codes, scanning, 16
University of Toronto Work and Health Research Institute, 40

Van Beek, John, 57, 121
van Eerd, Dwayne, 33, 42, 137
Vandelac, Louise, 200
VDT Solutions, 110–11
video coding, 167–8
Virginia, workers' compensation in, 86
virtual reality, 117
visual display terminals, 115–16, 165, 172–4. *See also* computers; keyboarding
voice recognition. *See* true voice recognition

W5 (television program), 131
wages, 202: rollbacks, 159; of women, 152–3
Walker, Cathy, 120, 147
Wands, Susan, 133
Washington State: ergonomics guidelines, 130; workers' compensation in, 7–8, 96
websites: ergonomics, 120; MSI,

58–9; reflex sympathetic dystrophy (RSD), 51; RSI, 39
Weininger, Lois, 151
welfare, 159, 192, 197–8
white-collar workers, 105–6, 150
Wilkomir, Richard, 108
Wilson, William Justin, 196
Wire Service Guild, 148
Witmer, Elizabeth, 89
women: automation and, 196–8; CTS in, 12, 66; ergonomics and, 123–4; hours of work, 204; inequality of, xiii; injured, 89–90; living on street, 198; MSIs and, xii–xiv, 17, 31–2, 62–3, 95–6; pain and, 98; part-time work, 157; RSI and, 26–8; tendinitis in, 66; in trade unions, 143, 152–8
Women and Work Research and Education Society, 127, 150–1
work: benefits, 191; clerical, xiii, 6; contract, 191, 192; deskilling of, 6; domestic, 196–7; factory, 16; hours of, 111–12, 147–8, 202–4; pacing of, 136–7; position changes, 112–13; regulation of employment practices, 201–2; social aspects, 200; temporary, 192; unpaid, 197; *See also* assembly-line work; jobs; workers
Work and Health Research Institute, 40
workers: blacklisting of, 89; blue-collar, 29, 65, 137; consultation with, 121–3; contingency, 200; cost of MSIs to, 69–70; employer/worker teams, 125; fraud by, 85–6; input from, 171; interviews with, 127–8; involvement of, 172; new, 21; part-time, 97–8, 155, 157, 191, 192; pink-collar, 196; pre-employment testing, 104; quality of life, 140–1; safety duties, 175; temporary, 200–1; treatment by corporations, 159; unionization of, 143; VDT, 172–4; white-collar, 105–6. *See also* trade unions
workers' compensation, xiii, xiv, 74–99; in Alberta, 77, 93–4; in Australia, 29; in British Columbia, 7, 98–9, 120–1; filing claims, 92–5; fraud, 82–6; history of, 6; in Ontario, 88–9; premiums, 10; profitability of, 91; reform of, 86–91; in the U.K., 167; in the U.S., 29, 86–8, 90–1; in Washington State, 7–8, 96
Workers' Health and Safety Agency (Ontario), 9
workers' health centres, 39, 151
Workers' Memorial Day, 146
Workers of America, 160
workfare, 197–8
workforce: costs, 68; effect of automation on, 194–6; inequality in, xiii; surplus labour in, 195
Working Today, 161
Work-Loss Data Institute, 66
workmen's compensation. *See* workers' compensation

Workplace Health and Safety Agency (WHSA), 78, 83

Workplace Safety and Insurance Board (Ontario), 77

workplaces: analysis of, 129–30; arrangement of space in, 107; centralized, 199–200; changes in, 13, 122–3; components of, 20; deaths in, 145–7; design of, 19–20; hazards in, 161–2; health regulations, 76–8; inspections, 165; safety regulations, 76–8; temporary workers in, 20

worksites. *See* workplaces

workstations, 100; adjustment of, 36; computer, 105–17; guidelines for, 130–1; platforms, 133

Wright, Graham D., 25–6

wrist position, 36–7, 108–10, 124, 169. *See also* carpal tunnel syndrome (CTS); hands

writer's cramp, 5–6

yoga, 44

Yukon: occupational health and safety committees, 135; workers' compensation in, 77

Z., Audrey, 11

*Zarecki* v. *Amtrak*, 104